SpringerBriefs in Law

More information about this series at http://www.springer.com/series/10164

Kiran Krishnan Nair

Small Satellites and Sustainable Development - Solutions in International Space Law

Springer

Kiran Krishnan Nair
Institute of Air and Space Law
McGill University
Montreal, QC, Canada

ISSN 2192-855X ISSN 2192-8568 (electronic)
SpringerBriefs in Law
ISBN 978-3-030-18619-7 ISBN 978-3-030-18620-3 (eBook)
https://doi.org/10.1007/978-3-030-18620-3

This Springer imprint is published by the registered company Springer Nature Switzerland AG.
The registered company address is: Gewerbestrasse 11, 6330 Cham, Switzerland

Acknowledgements

This endeavour has been a process of learning. A process, during which I learnt, unlearnt and relearnt a variety of aspects related to international law. My sincere thanks and deepest respect go to my supervisor, Prof. Ram S. Jakhu, who introduced, guided and enabled me on this process of learning. I am equally grateful to the Faculty of Law, McGill University, for granting me the Erin J. C. Arsenault Fellowship in Space Governance and Graduate Excellence Award in Law.

I would also like to place on record my gratitude to the staff and faculty of the Institute of Air and Space Law (IASL), in particular, the Director, Prof. Brian Havel, for his persistent support and encouragement. Sincere thanks also to Ms. Maria D'Amico, the 'angel' of IASL for her positive support. My gratitude also extends to Dr. David Kendall, Chairman, UNCOPUOS, for his kind indulgence in providing me first-hand insights into the role of international law in shaping international organisations. I am also indebted to Ms. Mary Lucy Stojak, Director, HEC, and Chair of the Canadian Space Advisory Board, for comprehensively reviewing this work and enriching it with her deep insight, patience and encouragement. Sincere thanks in no small measure to Ms. Pasqualina Chiarelli, Student Affairs Coordinator at McGill University, for her unstinting administrative support and assistance.

A note of thanks goes to my numerous friends and well-wishers at McGill who made the process of learning so much more enjoyable and wholesome. Towards this, I express my sincere gratitude to all my friends. I did attempt a listing of friends, but the list is too long to fit in and I am sure my friends would allow me that latitude.

My debt of gratitude is acknowledged in respect of my extended family of that wonderful body of personnel who constitute the Indian Armed Forces, and its fantastic institution—the Indian Air Force. And finally, the greatest debt is gratefully acknowledged in respect of my wife Sreeletha for her enduring patience and support as also my two little children, Rohtaksh and Mahaksh, for the endless welcome distractions.

My profound thanks to all of the above and my humble apologies to those I missed.

Acknowledgments

Contents

Abbreviations

AASL	Annals of Air and Space Law, McGill University
ABM	Anti-Ballistic Missile
ASAT	Anti-Satellite
BMD	Ballistic Missile Defense
CD	Conference on Disarmament
COPUOS	Committee on Peaceful Uses of Outer Space
CS	Constitution, International Telecommunication Union
CTBT	Comprehensive Test Ban Treaty
CV	Convention, International Telecommunication Union
DOC	Department of Commerce, United States
DOS	Department of State, United States
ECSL	European Centre for Space Law
ESA	European Space Agency
EU	European Union
GEO	Geostationary Earth Orbit
GSO	Geosynchronous Orbit
HEO	Highly Elliptical Orbit
ICAO	International Civil Aviation Organization
ICJ	International Court of Justice
IISL	International Institute of Space Law
ILC	International Law Commission
ISRO	Indian Space Research Organisation
ITU	International Telecommunication Union
LC	Liability Convention
LEO	Low Earth Orbit
MA	Moon Agreement
MEO	Medium Earth Orbit
NASA	National Aeronautics and Space Administration
OST	Outer Space Treaty
PAROS	Prevention of an Arms Race in Outer Space

RC	Registration Convention
SSA	Space Situational Awareness
UN	United Nations
UNGA	United Nations General Assembly
UNOOSA	United Nations Office of Outer Space Affairs
VCLT	Vienna Convention on the Law of Treaties
WW II	World War II

Chapter 1
Introduction

Abstract The year 2018 marks 50 years since the first United Nations Conference on the Exploration and Peaceful Use of Outer Space. The year shall pass and the world would change as it always has. The only constant would be change and the dynamism of law to respond to change and prevail. International space law has been alive to the shifting patterns of technology, geo-politics, geo-economics and has evolved to remain relevant. What remains to be seen is whether it would continue to be as relevant in view of the paradigm shift in the number and variety of actors and issues in space today. Accordingly, this book focusses on a review of traditional concepts in the changed context and explores solutions that are innovative, unconventional and yet contextual as also pragmatic.

Space activities grew out of the military conflict of WW-2 and the law pertaining to outer space evolved during the dramatic manifestations of military space competition amongst the two super powers. For well around four decades, the bulk of satellites were military, they were enshrouded in secrecy, civil space uses were few and the prime users were military forces. Secondly, possession of space assets was mostly confined to states, particularly the super powers, and private, commercial owners and uses were few. The stakeholders were largely military, national interests were divergent and hence competition rather than cooperation was the prime area of focus in space. The desire to forestall military competition in space and not regulation of commerce or civil development led nations to cooperate in the creation of international space law, in particular the Outer Space Treaty-1967.[1] Thus, the scope, objective and purpose of the law for well over four decades was driven by competition and international space law was focused largely on ensuring that earthly military competition did not spiral into outer space.

Today, the issue of earthly military conflict spiraling into space has not disappeared altogether. It continues to be relevant and yet, equally or more relevant is the fact that space activities are now increasingly civilian, dispersed and prolific in character. Unlike the past, out of the 1738 operational satellites in orbit, barely 20%

[1] Treaty on Principles Governing the Activities of States in the Exploration and Use of Outer Space, including the Moon and Other Celestial Bodies, 27 January 1967, 610 UNTS 205 (entered into force 10 October 1967) [OST].

K. K. Nair, *Small Satellites and Sustainable Development - Solutions in International Space Law*, SpringerBriefs in Law, https://doi.org/10.1007/978-3-030-18620-3_1

are military.[2] Also, space capabilities are no longer confined to the superpowers with over 66 countries operating satellites and over 70 space agencies in existence.[3] Space capabilities are no longer in the exclusive domain of states but are common place with satellites being used for a variety of tasks ranging from banking, weather, agriculture, hydrology, town-planning to telemedicine and tele-education. Space uses and users expand rapidly considering that the number of civilian users of military GPS today far exceed their original military users. Space tourism, so relevant today was not even contemplated when the OST was formed. Nano satellites were unknown in the past and as of January 2018, there are 560 nanosats in orbit[4] and constellations of thousands are planned in the next few years.

Apart from space objects, the actors in space are also more diverse today as also the models and means of cooperation. There exists convergence and cooperation amongst states, states and non-state actors, within corporations and also between companies and individuals on a wide variety of areas ranging from sharing of costs to technology exchange, civil development. There is also cooperation and convergence among civil and military uses of space. The picture is fluid and common factor binding this varied kaleidoscope in space today is common interest. It enables mutual benefit by sharing resources and cost in addition to providing a means of tackling common vulnerabilities. Thus, one is now witness to an increased clamor for collaboration on areas of common vulnerability like space debris, collision avoidance, space traffic management, and orbital slot and spectrum management. These areas have today gained far more prominence than ever before and would be increasingly critical in the future. In the past, states have cooperated to contain competition in space, the need for the present and the future is to cooperate simply because it is mutually beneficial, enables optimal utilization and serves the common good of most, if not all.

On the other hand, the envelope of space activity has widened and now spills way beyond the original and intended regulatory confines of the specialized treaty, like the OST. For instance, the question of space debris not specifically addressed by the OST is covered to an extent by international environment law. Similarly, the aspect of 'colonization' of planets and appropriation of celestial bodies demand application of a variety of principles of international law ranging from bio-ethics to international property. Hitherto unknown issues like mega-constellations of small satellites would seek answers in the theoretical precepts of international tort law, the law of international contracts and international liability amongst other since the activity by its very nature is international in character. In the changed circumstances, it would be essential to explore if and to the extent to which the public international law principle of '*Rebus Sic Stantibus*' applies and the attendant remedies. International law and the space treaties particularly lay great emphasis on promoting international

[2] Union of Concerned Scientists, "UCS Satellite Database", online: https://www.ucsusa.org/nuclear-weapons/space-weapons/satellite-database#.WptYTOjwbIV (accessed on 10 April 2017).

[3] Space Security Index, "Space Security Index 2016" (September 2016), at 55.

[4] Erik Kulu, *Nano Satellite Database*, online: http://www.nanosats.eu/index.html#info (accessed on 10 April 2017).

cooperation in the conduct of space affairs which is an aspect that is no longer programmatic but is pragmatic. The role of international law in regulating issues in space becomes increasingly critical as the number of actors and issues keep rising beyond what was envisaged.

This paradigm shift in space has been drastic, unforeseen and demands legal focus and regulation. However, most legal literature continues to be largely engaged with the manifestations of military competition in space. This perhaps is explained by the fact for around four decades, the military manifestation were profoundly dramatic and unnerving. The period from the late 1950–60s was characterized by rapid deployment of military satellites followed by anti-satellite and nuclear tests in space in the following two decades. The 1980s were witness to the Strategic Defence Initiative or "*Star-Wars*" that continues to influence popular imagination till date. The 1990s were no different with the Gulf war being called the "first Space war" because it was for the first time such a large number of space objects were brought to bear on war. Even the 2007 Chinese ASAT test that littered debris was a source of great attention. Consequently, the focus of space literature is largely on three aspects:

Firstly, the early military competition and manifestation of legal regimes.[5]
Secondly, controversies on legal aspects, particularly definitions and implications.[6]
Lastly, questions on the legality of military uses of space and legal reform to reverse militarization.[7]

In the latter theme two opposing camps are clearly evident. One supports binding legal framework and the other favors 'soft-law' options like a space code-of-conduct to almost all issues, including the desire to reverse space militarization. In keeping with the dramatic manifestations of military competition, for well around four decades, literature on space has mostly focused on variations of the above themes. The third theme in particular abounds in literature and yet no legislative reform has taken place since the Moon treaty of 1979; i.e. the traditional narrative of stalling military competition by introducing new legislation continues to dominate. However, evidence indicates that such legislative endeavors yield little, if at all anything. For instance, the last treaty preventing military activity in space was made in 1967, the UNCOPOUS[8] has been deadlocked since 1995, the UN Conference on Disarmament has been considering proposals since the early 1980s to no avail[9] and soft-law approaches proposed are yet to make any headway.

---

[5]Krepon (2004) and Hitchens (2007).

[6]Benko et al. (2005) and Jakhu et al. (2016).

[7]Hays (2011).

[8]The mandate of UNCOPUOS is limited: "to review the scope of international cooperation in peaceful uses of outer space, to devise programmes in this field to be undertaken under United Nations auspices, to encourage continued research and the dissemination of information on outer space matters, and to study legal problems arising from the exploration of outer space". *International Co-operation in the Peaceful Uses of Outer Space*, Res 1472 (XIV) UNGA, GAOR, 14th Sess., Supp. No. 16, A/PV.857 (12 December 1959).

[9]United Nations Office of Disarmament Affairs, *Outer Space: Efforts by the Conference on Disarmament*, online: https://www.un.org/disarmament/topics/outerspace/ (accessed on 15 April 2017).

By contrast, convergence and cooperation is visible among the various actors and issues in space and yet legal literature on these aspects is scant. In particular, the role of international law in regulating the wide variety of new issues and actors in space has not been dealt with. The available literature is mostly episodical and limited to news reports, reports of the UNOOSA,[10] which serves as the Secretariat of the UNCOPUOS and studies[11] that deal in general terms with the subject.

This thesis seeks to explore the specific issue of the extent to which the incentives for cooperation rather than competition have increased and their impact, what existing international legislation applies to the new areas of commerce and civil development, to what extent it applies, to what extent it encourages or inhibits the spurt in civil cooperation and commerce, whether existing international legislation can be gainfully adapted to the changed situation, which areas are readily adaptable and which require modification or change in entirety. Legal review and reform for military activity is stagnant because of legacy issues, but the emerging areas of civil convergence, cooperation are novel and hence may be amenable to legal review and reform. The rationale is not to take the usual recourse of using law to proscribe military activity but to lawfully employ regulations and principles of international law for promotion of common interests.

The changed context provides the right opportunity to shift focus from containing conflict to encouraging convergence. An examination of the recent surge in civil and commercial activities as opposed to military activity would be undertaken followed by an exploration of international legislation enveloping space issues to ascertain the scope, objective and purpose of legislation in the changed context. A comparative analysis of domestic space legislation with a particular focus on regulations states have enacted recently for cooperation, commerce and governance would also be undertaken. Civil aspirations common to states would be identified and the extent to which existing provisions of international law fulfill aspirations and remedial measures possible would be explored. A mixed method that relies primarily on doctrinal analysis that is qualitative and yet uses quantitative research methods to support conclusions is proposed. Methodical triangulation involving use of data of various state practices, theories and analogies would be used to validate inferences.

In the changed context, the need for better governance for using scarce space resources, commerce, environmental and space security has assumed primacy. Almost all nations use space and aspire to better use of space resources leading to increased instances of technology transfer, resource allocation and cost distribution. This upsurge in international cooperation amongst a variety of state and non-state entities for civil use and commercial gain drastically widens the context. The context is no longer confined to two superpowers vying for military supremacy but is far more diffused in terms of players and scope. In such a wide context, the scope of application of the principles of international cooperation and development contained in the space treaties is far wider. A focus on the affirmative obligations rather than negative proscriptions may be more beneficial. The change in focus explores new

[10] UNOOSA (2016).

[11] Jakhu and Pelton (2017).

areas making the research unique and academically relevant. This research seeks to use the changed paradigms to explore utility of existing international legislation from an altogether new perspective. This thesis argues that while the international law has remained unchanged, the situation has changed. The changed situation presents new opportunities that this exploratory endeavor seeks to pursue.

References

Benko M et al (2005) Space law: current problems and perspectives for future regulation. Eleven International Publishing, The Hague

Hays PL (2011) Space and security: a reference handbook. ABC-CLIO, Santa Barbara

Hitchens T (2007) Debris, traffic management and weaponization. Brown J World Aff 14(1): 173

Jakhu R, Pelton J (2017) Global space governance: an international study. Springer, New York

Jakhu R, Steer C, Chan, K-W (2016) Conflicts in space and the rule of law. Retrieved from https://papers.ssrn.com/sol3/papers.cfm?abstract_id=2722245

Krepon (2004) Space assurance or space weapons? Georget J Int Aff 5(2): 3

UNOOSA (2016) Contribution of space law and policy to space governance and space security. In: 10th UN workshop on space law, Vienna, 5–8 Sept 2016

Chapter 2
The Principle of Peaceful Use of Outer Space: Reviewing the Scope of "*Peaceful*" in the Changed Context

Abstract As we move into the future, the role of law in maintaining peace in space continues to be relevant, but equally relevant is the role of law in enabling peaceful use of space as is ordinarily understood. A peaceful use not just restricted to conflict prevention, but the wider expanse of conducting affairs in space 'peaceably', without fear of debris collisions, environmental degradation and unfair commercial gain, competition among other things. Towards this end, this chapter seeks to go beyond the traditional narrow confines of the treaty interpretation of 'peaceful use of space' to a broader, more contextual and pragmatic interpretation.

2.1 The Enigma of Peaceful Use of Space

Technology enables humankind to adapt to the unnatural environment of sea, air and space. Adaptation is followed by competition and hence it comes as no surprise that specialized legislation exists to regulate affairs in the distinct environments of sea, air and space. However, what is intriguing is that despite sea and air being the scene of pitched battles it is only in space legislation that one comes across an overwhelming emphasis of the term 'peaceful use of outer space'. The *Lex Specialis* of space, the corpus of Outer Space Treaties and principles evolved during the intense military competition of the Cold-War and so deep is the association of the term 'peaceful' with conflict that the corpus explicitly mention the term thirty-six times.[1] By contrast, the terms 'commercial use' and 'civil use' find no mention in the space treaties and principles although today commercial and civil uses of space far exceed its military uses.

[1]Treaty on Principles Governing the Activities of States in the Exploration and Use of Outer Space, including the Moon and Other Celestial Bodies, 27 January 1967, 610 UNTS 205 (entered into force 10 October 1967) [OST].

Union of Concerned Scientists, "UCS Satellite Database", online: https://www.ucsusa.org/nuclear-weapons/space-weapons/satellite-database#.WptYTOjwbIV (accessed on 10 April 2017).

K. K. Nair, *Small Satellites and Sustainable Development - Solutions in International Space Law*, SpringerBriefs in Law, https://doi.org/10.1007/978-3-030-18620-3_2

So powerful has been the impact of the principle of 'peaceful use of space' that entire regimes and institutions like the United Nations Committee on Peaceful Uses of Outer Space (UNCOPUOS) crystallized around it. The Committee was created by the United Nations 59 years ago in 1959 with 18 states and now comprises of 87-member States with three more applications to be approved in 2018 and 37 inter-governmental and non-governmental international organizations as permanent observers, with an accelerating increase in membership over the past several years (a 30% increase in membership since 2010).[2] In keeping with the intense military competition then prevailing, it's original mandate was to ensure that terrestrial conflict did not spiral into space though it has now broadened to include strengthening of the international space regime governing space activities as well as building a collaborative approach to address vulnerabilities and sustainability of space activities. This institutional broadening of the scope of peaceful use is indicative of state practice expanding beyond a narrow spectrum of military use to a wider expanse of civil and commercial use. The widening of institutional arrangements dealing with peaceful use indicates an implicit widening of the original intention of what the notion of peaceful use entails.

Despite the above, most writing on the principle of peaceful use has also been against the narrow military backdrop of 'aggressive or non-aggressive use' and not in the broader context of a trouble-free environment that peaceably allows people to go about their business. Consequently, it would be in order to ascertain the causes that led to such preponderance, its relevance to the present and future and whether it now makes sense to unshackle the principle of peaceful use of outer space from its military moorings and interpret it in its normative wide context. Towards this end, recourse would be taken of the Vienna Convention on the Law of Treaties (VCLT).[3] Though strictly speaking not applicable to the OST since it came into force in May 1969 after conclusion of the OST, the rules of customary international law codified in the VCLT do apply to the OST. Pursuant to the main rule of treaty interpretation laid down in the Vienna Convention, Article 31 (1):

> A treaty shall be interpreted in good faith in accordance with the ordinary meaning to be given to the terms of the treaty in their context and in the light of its object and purpose.[4]

Apparently, the ordinary meaning is to be construed so as to be contextually applicable which dispenses with any notion of a rigid adherence to a particular interpretation. This work shall explore the contextual applicability of the existing interpretation of 'peaceful use of space' to the present and future. In doing so, the first step would be a brief exploration of the evolution of the concept.

---

[2] Kendall (2018).

[3] *Vienna Convention on the Law of Treaties*, 23 May 1969, 1155 UNTS 331, 8 ILM 679 (1969), 63 AJIL 875 (1969) (entered into force 27 January 1980) [VCLT].

[4] *Ibid.* Article 31.

2.2 Exploring Evolution of the Concept of Peaceful Use of Outer Space

The first use of space manifest in the launch of Sputnik was peaceful by today's standards in that it was non-aggressive and did little apart from beeping down signals to earth. However, the 'Sputnik Shock'[5] disturbed peace and managed to cause a wave of insecurity across the US in that the launch symbolized the ability of the Soviet Union to undertake military observation and weapon delivery over the US utterly unhindered.[6] It would be in order to explore in some detail whether the reverberations of Sputnik were the sole cause of the existing interpretation of peaceful use or whether other factors also contributed significantly and the same is attempted below.

2.2.1 Peaceful Use of Space in the Pre-Sputnik Era

As in case of Air law, legislative endeavors on space preceded the actual technology involved and while the Sputnik was launched on 4 October 1957, the origins of space law can be traced back to as early as 1910 when a Belgian lawyer, Emile Laude wrote on the need for a new law to govern judicial relations in space.[7] The definitional aspect of where the boundary of air space ends and outer space begins were discussed as early as 1926 by V. Zarzar, a Soviet official. Later, in 1934[8] Eygene Korovin, professor of international law in Moscow, published his article on the international legal aspects of the stratosphere, "*La conquête de la stratosphere et le droit international*".[9] Many more such writings on a variety of abstract concepts in space followed, but what remained consistently absent was the question of "*peaceful uses of outer space*". In fact, the term was absent in the space legal literature of the Soviets, the US, UK and other European powers. Since it finds no place in space literature of those times it may be safely assumed that the term was drawn from analogous treaties which formed the sub-strata of later space treaties.

Historical legal analogies usually form the basis of fresh law formulation and hence to fill the vacuum of space law most of the initial efforts at creating laws in outer space were premised upon three analogies—air, high seas, and Antarctica.

---

[5]For details, see "News-Conference Remarks by President Eisenhower, 9 October 1957" in Bess CM Reijnen, *The United Nations Space Treaties Analysed* (Utrecht: Editions Frontieres 1992), at 26–37.

[6]Dickson (2001).

[7]Laude (2002) and Doyle (2002a), at 1.

[8]The Soviets were the early pioneers of space law. For example, Dr Stephen E. Doyle in his book "Origins of International Space Law" refers to papers presented at an air law conference in Moscow in 1926 that dealt with debates about the nature and extent of national sovereignty in airspace. Additionally, mention is also made of Czech professor Vladmir Mandall writing the world's first major work of substance on the subject of space law in 1932. See Laude *supra* note 7.

[9]Korovin (1934).

Each of these analogies suggested a distinct approach to the regulation of space. The air and high seas analogies implied treating outer space as open to forms of military activity accepted under general international law, while the Antarctic analogy suggested treating outer space as 'off limits' for all military activity. The above three analogies are briefly dwelt upon below so as to appreciate and examine its conceptual applicability and comprehend how they found their place in the later space treaties.

The Air Analogy: The air analogy supported notions of state sovereignty and control over all activity above its territory. It implied that the same rules regarding military activity that prevailed within a state's own domain, including its airspace, should be applied to outer space. These rules included the right to construct and maintain weapons and armed forces, and to use armed force against unauthorized intruders in self-defense. For example, military aircraft intruding upon national airspace can be shot down whereas civilian aircraft can be escorted or forced down as the case may be.[10] Prior to 1932, there existed consensus that sovereignty must prevail in air but not space with Laude (1910),[11] Zarzar (1926)[12] and Mandl (1932) conceptually asserting that flight in ether or space should be free and unconstrained.[13] Towards 1934, one witnesses dissent in the works of Soviet scholars like Korovin who asserted sovereignty should extend to space. Regardless of the outcome of these debates, the sovereignty aspect of air was not applied and space overflight of national territory is an accepted norm today.

The Sea Analogy: The basic rationale for this principle was set forth by the Dutch Jurist, Hugo Grotius (who eventually became known as "the father of international law") in 1609 in his famous *Mare Liberum*.[14] This analogy supported the notion of freedom of the seas. Freedom of the seas is the principle that outside its territorial waters, a state may not claim sovereignty over the seas except with respect to its own vessels. He came up with a concept that translated to mean '*Free Seas*'. It meant that the sea should be open to all nations.[15] The seas cannot be appropriated by one sovereign, or even by a number of them. It was from the above that concepts related to the '*common heritage of mankind*' were derived.

The Antarctic Analogy: This analogy available after completion of the Antarctic Treaty, suggested non-militarization of an entire area. The treaty stated that Antarctica shall be used "*for peaceful purposes only*,"[16] and defined this to mean a prohibition on all military activities, inclusive of

- The establishment of military bases and fortifications.
- Carrying out military maneuvers.
- Testing any type of weapons.

---

[10] Foont (2007), 695 at 718.

[11] As noted by Jenks (1965), at 97.

[12] Doyle (2002b), at 1–4, quoting extracts from V. A. Zarzar.

[13] Mandl (1932).

[14] Grotius (2004).

[15] Almond (1991).

[16] *The Antarctic Treaty*, 1 December 1959, 402 UNTS 71 (entered into force 1961).

This entailed far more comprehensive limitations than those which prevailed within the state domain or on the high seas. It banned even forms of military activity regarded as defensive under the UN Charter. Nevertheless, it allowed for military personnel and equipment to be used for scientific research and for peaceful purposes. Essentially, it enabled a compromise by keeping the environment free of all military activity on one hand and on the other drawing on the advanced scientific competencies of military personnel and equipment. To a world recovering from the ravages of World-War-II, the principles of the Antarctic treaty made eminent sense and hence many aspects of the concept of peaceful use of the Antarctic found insertion almost verbatim into the first space treaty.

2.2.2 *Crystallization of the Concept of Peaceful Use of Outer Space*

The memories of World War-II were fresh and hence prior to even the launch of the Sputnik, the world community inclusive of the then prevailing space powers of US and Soviet Union overwhelmingly favored the use of space for peaceful purposes. In the circumstances then prevailing, peaceful use clearly meant no military use whatsoever. Further moves to ensure that 'outer space be used exclusively for peaceful and scientific purposes and for the benefit of mankind' included the joint submission by four Western powers (Canada, France, the United Kingdom and the United States) to the United Nations Disarmament Commission, calling for a study on an inspection system that would assure that objects launched into outer space would be used exclusively for peaceful and scientific purposes.[17] Adopted by the General Assembly, this became the first United Nations resolution on outer space, and the first time the phrase 'exclusively for peaceful purposes' would be used in an authoritative United Nations text.[18]

However, the intent for peaceful uses would not have evolved beyond platitudes but for the launch of Sputnik. The launch of Sputnik was not an isolated event and since international law is strongly influenced by the contemporary geopolitics, it should be seen in the sequence of events relevant to the geopolitics then prevailing. Prior to Sputnik, in August 1953, the Soviets tested their Hydrogen bomb and on 3 August 1957, the Soviets successfully flight-tested the world's first ICBM, the R-7 codenamed SS-6 Sapwood.[19] Based on their SS-6 ICBM booster, the Soviets on 4 October 1957, launched the world's first artificial satellite-*Sputnik-1* (Traveller-1), thus heralding in the dawn of the space age.[20] With the above launch, the superiority

[17] *Western Working Paper Submitted to the Disarmament Subcommittee: Proposals for Partial Measures of Disarmament*, UN Doc DC/SC.1/66 in *Fifth Report of the Sub-Committee of the Disarmament Commission* UN Doc DC/113, Annex 5, 869 at 871.

[18] *Question of the Peaceful Use of Outer Space*, Res 1348 (XII), UNGA, 13th Session (14 November 1958).

[19] Peebles (1983), at 52.

[20] NASA, *Sputnik and the Dawn of the Space Age*, online: https://history.nasa.gov/sputnik/.

of Soviet military space technology was conclusively demonstrated and it was evident to the entire world that the Soviets now possessed the powerful military troika of nuclear weapons, ICBMs and satellite launchers. Soviet morale sky-rocketed while US national morale nose-dived. The Soviets added insult to injury by offering assistance to the US through the UN program for technological assistance to primitive nations.[21] The Soviet capacity to use space for delivery of munitions as well as spying on the US caused enormous uproar in the US.[22] Subsequent to the Sputnik shock, the Americans, to counter the Soviet military advantage in space entrusted the responsibility for space to its Air Force.[23] Post-Sputnik, the world had changed; it was evident that the use of space 'exclusively for peaceful purposes' was seriously questionable. To the contrary, it was increasingly becoming a domain for military competition.

Thus, it came as little surprise that the thirteenth session of the General Assembly, held in 1958, began seriously debating 'Questions of the Peaceful Use of Outer Space'. The item relating to the peaceful use was included for the first time and during this session the term 'peaceful' was used as an antonym to 'military'. The General Assembly adopted resolution 1348 (XIII), which recognized the 'common aim' of humankind that outer space 'should be used for peaceful purposes only.'[24] The world was deeply concerned and nations like Sweden appealed to fellow Member States to 'safeguard outer space against any military use whatsoever'.[25] The concerns evolved beyond paper and hortatory declarations to institutes and infrastructure. Institutional arrangements began to coalesce around the concept of peaceful use with the thirteenth session establishing the Ad Hoc Committee on the Peaceful Uses of Outer Space. The following year, at its fourteenth session, the General Assembly by resolution 1472 A (XIV) established a permanent body, the Committee on the Peaceful Uses of Outer Space (COPUOS).[26] Starting in 1958, the General Assembly passed a number of resolutions establishing basic concepts for a space law regime.[27] These concepts include: that international law, including the UN Charter, is applicable to outer space and celestial bodies; that outer space and celestial bodies are free for exploration and free from national appropriation; that principles such as state responsibility, ownership, and control be applied to the operation of space vehicles; and that arms control principles are applicable to space. However, resolutions being non-binding have limitations and had but little effect on the military race in space.

---

[21] Reed (1987).

[22] Roger D. Lanius, "Sputnik and the Origins of the Space Age" online: https://history.nasa.gov/sputnik/sputorig.html, accessed on 18 May 2017.

[23] By 1961, the US Department of Defense had entrusted the entire mission of managing and operating US military space launch vehicles and satellites to the Air Force.

[24] General Assembly Res 1348, *supra* note 29.

[25] McDougal et al. (1963), at 395.

[26] *International Cooperation in the Peaceful Uses of Outer Space*, Resolution 1472 (XIV), UN General Assembly, 14th Session (12 December 1959).

[27] UN General Assembly Res 1348 (XII), *supra*, note 29.

2.3 Convolution of the Concept of Peaceful Use of Outer Space

The first few years following Sputnik were witness to a flurry of intense military space activity. Despite lofty declarations of peace by both the US and Soviet President in 1958,[28] the superpowers rapidly launched military satellites to the extent that within the first four years itself, almost the entire range of satellites possible for military use were in orbit (see Table 2.1).[29] As the number of military satellites grew, the means to destroy them also rose and a number of Anti-SATellite (ASAT) tests were conducted by both the US and Soviet Union.[30] This also included nuclear ASAT tests in space with the US conducting up to six nuclear tests in space and the Soviets also doing likewise (see Table 2.2).[31] The scenario in space quite clearly revolved around the superpowers and the arms race in outer space was both very dramatic and disconcerting to the world at large. As borne by Jasentuliyana:

> During the 1960s, the USSR and US were dominant in spaceflight activities. For subjects on which these two powers could agree, it was possible for the UN to formulate and obtain general assent to international agreements relating to spaceflight activities.[32]

There was but little the super powers could then agree upon and this included the futility of the use of nuclear weapons in space. Nuclear weapons caused indiscriminate damage in space and served little military purpose for both players in space and consequently, the first breakthrough in stalling the arms race in space came with the Partial Test Ban Treaty (PTBT) of 1963.[33]

2.3.1 *The Genesis of Selective Interpretation of Peaceful Use of Space*

The PTBT was motivated, in part, by the fact that the Electro Magnetic Pulse generated by the US and Soviet nuclear tests in space disabled at least six satellites of the US, Soviet Union and also UK.[34] It was clear to the superpowers that nuclear ASAT tests in space led to indiscriminate damage to all and hence it made sense in stopping such tests in space. This realization also enabled adoption of UNGA Res 1884 (XVIII) in the same year which called on states to refrain from placing in orbit

---

[28] Christol (1991a), at 15–16.

[29] See Table 2.1 A, *infra* at 36.

[30] Grego (2012).

[31] See Table 2.2, *infra*, at 37.

[32] Jasentuliyana (2010), at 8.

[33] *Treaty Banning Nuclear Weapon Tests in the Atmosphere, in Outer Space, and Under Water*, 5 August 1963, 480 UNTS 43; 1964 UKTS 3, Cmnd 2245; 14 UST 1313, TIAS 5433; 1964 2 ILM 883 (entered into force 10 October 1963), at 889–891.

[34] Alves (1991), at 70.

Table 2.1 First satellites in orbit[a]

Satellite	Number of satellites-year wise										First Soviet launch date
	1958		1959		1960		1961		Total		
	US	USSR	US	USSR	US	USSR	US	USSR	US	USSR	
Communication	01	00	00	00	02	00	00	00	03	00	1964
Navigation	00	00	01	00	01	00	03	00	05	00	1967
Photo-recce	00	00	06	00	06	00	13	01	25	01	1962
Early warning	00	00	00	00	02	00	03	00	05	00	1971
Meteorology	00	00	00	00	02	00	01	00	03	00	1963

[a]Data on satellites has been collated from various sources including the UNOOSA On line Index of Space Objects online: http://www.unoosa.org/oosa/osoindex/search-ng.jspx?lf_id; Gunter's Space Page, satellite listing online: http://space.skyrocket.de/directories/sat.htm; and the satellite data base of Union of Concerned Scientists, online: https://www.ucsusa.org/nuclear-weapons/space-weapons/satellite-database#.XA-Tqid7nOQ

Table 2.2 Nuclear tests in outer space[a]

US nuclear tests in space				Soviet nuclear tests in space			
Test	Date	ALT (km)	Yield (kt)	Test	Date	ALT (km)	Yield (kt)
ARGUS-1	27 Aug 1958	200	1.5	K-project (127 K-2)	27 Oct 1961	150	1.2
ARGUS-2	30 Aug 1958	240	1.5	K-project (127 K-1)	27 Oct 1961	300	1.2
ARGUS-3	06 Sep 1958	540	1.5	K-project (184 K-3)	22 Oct 1962	290	300
Starfish	09 Jul 1962	399	1400	K-project (187 K-2)	28 Oct 1962	150	300
Check mate	20 Oct 1962	147	20				

[a]Data collated from Nils-Olov Bergkvist and Ragnhild Fern "Nuclear Explosions from 1945 to 1998" *Report of FOA and SIPRI* at http://www.iaea.org/inis/collection/NCLCollectionStore/_Public/31/060/31060372.pdf and site of Arms Control Association at https://www.armscontrol.org/factsheets/nucleartesttally

around earth any object carrying nuclear weapons or any other kinds of weapons of mass destruction, installing such weapons on celestial bodies or stationing such weapons in outer space in any other manner. This text later found insertion verbatim in Article IV(I) of the OST. On the other hand, Ballistic Missiles supported a variety of strategic purposes for both sides. Additionally, transit of BMs through space caused no indiscriminate damage and hence both sides saw no sense in forestalling competition. Thus, no international legislation on stopping the transit of BMs through space came about. In effect, BMs, in the convoluted interpretation then prevailing constituted peaceful use of space. Likewise, an ingenious interpretation was extended to military satellites also so that its use was also constituted to be peaceful use. The evolution of the interpretation is best expressed by a CIA release in year 2000 of the US position in the 1962 UNCOPUOS meeting which identifies the problem as:

> There exists a possibility that a neutral or bloc member of the Committee will propose that the Committee, or its Legal Subcommittee, draw up a definition of "peaceful uses" or "peaceful purposes" in the outer space context. One purpose of such a proposal would probably be to have the Committee, and possibly the general Assembly, adopt a resolution restricting the exploration and use of outer space to activities falling within the definition. A further

> purpose, regardless of the adoption of a resolution, could be to provide a basis for attacks by states upon various outer space activities of other states.[35]

The aura of mutual suspicion and competition is demonstratively apparent in the above problem statement as also the instructions which state:

> In this regard, the Delegation should resist any attempt to substitute the terms "non-military" for "peaceful" and "military" for "non-peaceful" in characterizing space activities. The conduct of activities in space by the military does not in itself make them aggressive in character. Under the terms of the UN Charter, for example, the use of satellites in meteorology and for communications, navigation, early warning and observation is entirely permissible.[36]

Thus, the conceptualization and interpretation of peaceful had to be balanced with the prevailing realities of those times. There were interests to be protected and advanced. In fact, the *travaux préparatoires* of the OST reveals that during the negotiations India had proposed to extend the application of 'exclusively for peaceful purposes' in the second paragraph of Article IV of the treaty to all areas of outer space.[37] This proposal was, however, rejected because neither the US nor the Soviet Union wished a final definition of 'peaceful uses' in light of the expected limitations this could have meant for both states in their future uses of outer space. Consequently, the interpretation of peaceful was stretched to include ongoing military activities of the superpowers like military satellites, BMs, and conventional ASAT tests. There was no way the clock could be turned back and the law evolved around the prevailing realities of those times. It applied eminently to the context then prevailing.

2.3.2 Geo-Politics and Regime Formation in Space

The process of regime formation in space was born out of the prevailing geo-politics of the age and played a clear role in the interest development of the prime actors then. Both the superpowers wanted to retain their technological, military edge and at the same time understood that control of the vastness of space was not possible and hence while leaving space free from appropriation, they retained the right to overfly other territories and also conduct conventional ASAT tests while banning nuclear ASAT tests.[38] The language of the Resolutions was normative but left enough space to safeguard the interests of the super powers and hence issues like the definition of "peaceful use" was left open and wide. The broad contours of what constituted peaceful use not in normal terms but in the specific context of what suited the prime actors and issues in space then had already taken shape by 1962. This compromise with reality, it could be surmised led to a tacit acceptance by all states of the circumstances then prevailing. This is evidenced in the first, most significant of UN space

[35] US Central Intelligence Agency (1962).

[36] *Ibid.*

[37] *UNCOPUOS Verbatim Records*, United Nations Doc A/AC.105/PV.3 (March 1962), at 63.

[38] Reijnen (1992), at 103–105.

instruments, the Declaration of Legal Principles Governing the Activities of States in the Exploration and Use of Outer Space which was adopted unanimously by the UNGA in 1963.[39] The specific institution created for peaceful use in space, UNCOPUOS then proceeded to elaborate the five space law treaties on this declaration.[40] The influence of the superpowers can be assessed considering that the argumentation presented by the delegates to the UNCOPUOS was centred solely on the drafts presented by the US and USSR to reach the desired unanimity. Despite other provisions put forward by various countries, the debate was dominated by drafts submitted by the super powers. It is clear from discussions during the COPUOS sessions that the ultimate purpose of establishing an international binding regime in space was solely to maintain space for peaceful use as interpreted by the superpowers.[41]

The OST concluded in December 1966 was in large part composed of passages from the GA Resolution with little or no change. For instance, the treaty by bare repetition of paragraphs of GA Resolution 1884-XVIII and 1962-XVIII, does not elucidate them or make them more precise, also, the few novel provisions are drafted loosely enough to allow multiple interpretations. Apart from the lack of provisions for authoritative interpretation there exist no provisions for sanctions in case of non-observance of the Treaty. This led to expert remarks in those times like:

> In the Outer Space Treaty we have then a rigidly contractual instrument, in essence a bilateral arrangement between the principal space-users.[42]

In hindsight, remarks like the above made sense then because apart from the superpowers, most other states had but an academic or notional interest with no real space objects in orbit and hence had little to lose. Commercial and private players had even lesser stake. As a matter of fact, even after the OST came into being in 1967, the prime players continued to be the superpowers until the Soviet Union collapsed in 1991. It could be argued that the adoption of the treaties was less a demonstration of diplomatic success in forging unanimity or international cooperation amongst states and was more reflective of the passing interest of states. The stakes were then smaller for others, stakeholders were far lesser and space was simply not as pervasive as it is today to matter to the world at large. For example, an interruption in satellite communication or TV broadcast today would concern or rather alarm a far larger population on earth today than an F-15 ASAT test of US or Soviet co orbital ASAT

---

[39] *Declaration of Legal Principles Governing the Activities of States in the Exploration and Use of Outer Space*, GA Resolution 1962 (XVIII) UNGA, 18th Sess., UN Doc A/RES/18/1962 (13 December 1963).

[40] These treaties include, *Treaty on Principles Governing the Activities of States in the Exploration and Use of Outer Space, including the Moon and Other Celestial Bodies*, 27 January 1967, 610 UNTS 205 [OST]; *Agreement on the Rescue of Astronauts and the Return of Objects Launched in Outer Space*, 22 April 1968, 672 UNTS 119 [ARRA]; *Convention on International Liability for Damage Caused by Space Objects*, 29 March 1972, 961 UNTS 187 [LC]; *Convention on Registration of Objects Launched into Outer Space*, 14 January 1975, 1023 UNTS 15 (in force 15 September 1956) [Reg Conv]; and *Agreement Governing the Activities of States on the Moon and Other Celestial Bodies,* UN Doc. A/34/664, New York, 5 December 1979 [Moon Agreement].

[41] For details, see Christol (1991b), at 311–312.

[42] Fawcett (1968), at 15–16.

test during the 1980s. The circumstances today are different and the need for peaceful use of space has expanded drastically in terms of both actors and issues. However, the interpretation of peaceful use has not concurrently changed ever since its inclusion in the space treaties. In real terms, it has served its primary purpose of ensuring the absence of military conflict in space. The environment has remained free of actual military conflict despite over fifty-four ASAT tests[43] and the Treaty has never been grossly violated.

2.4 Adapting Peaceful Use to Contemporary Issues

2.4.1 Intentional Ambiguity and Its Incorporation in Space Treaties

As observed by Jakhu, intentional ambiguity was incorporated into the treaties to ensure they were flexible and adaptable to change.[44] The drafters of the OST intentionally kept its scope broad enough to govern all future space activities. Therefore, the Treaty not only contains fundamental legal principles but also the guiding philosophy for the governance of outer space. Viewed from the above perspective, the theory of intentional ambiguity makes sense and is purposeful in addressing the various issues of modernity in space law. The treaty was drafted in an atmosphere of mistrust and military competition to reconcile the competing interests of superpowers and to account for military uses not conceived then but which could be covered under the rubric of peaceful use. The idea was to contain war from spiraling into space.

Regardless of the origins, nothing stops one from interpreting peaceful use as is normatively meant and nothing inhibits expansion of its scope, objective and purpose beyond military issues. There is no way the drafters could have envisaged present issues of space tourism or a surge in constellations of small satellites. And yet, many argue that space treaties have become outdated and that the public international law principle of *rebus sic stantibus* applies.[45] As per the latter, treaties shall become inapplicable when a fundamental change in circumstances take place. At first glance, a fundamental change in circumstances appears to have taken place and yet, if one scratches the surface, it becomes apparent that no drastic change has taken place in space. The circumstances have not changed, they have expanded to such an extent that they no longer fully resemble the past. The issues today are far more diverse, but this is no way indicates that the military issue have disappeared altogether. Going by the trends, they continue to be as relevant today, the military competition continues with the Soviet Union being replaced by Russia and China. France, India, UK,

[43]Grego, *supra* note 29 at 7.

[44]Jakhu (2005).

[45]Foust (2017), Taft (2017), 362 at 364–366 and Englehart (2008).

Israel, Japan and many more states boast of military satellites and other capabilities. Far more nations aspire to military space capabilities today than ever before. The circumstances have not changed for *res sic stantibus* to apply; only the number of actors has expanded. Similarly, the issues have expanded beyond conventional military affairs to new issues of multiple stakeholders, small satellites, debris, and congestion amongst others. The context has widened and today includes a spectrum of peaceful use including in its expanse military activities on one end to commercial regulation, tourism on the other end.

2.4.2 Contextual Assessment of the Concept of Peaceful Use of Space

The notion of interpreting peaceful entirely against a military background is also a narrow construct of the previous era. The normative interpretation of peaceful is wider and means being free from disturbance, untroubled and calm. Following a full-fledged appraisal of the etymological origins of term 'peace' and related words in other languages, Prof Roger W Wescott arrives at the conclusion that the English term 'peace' draws on the Greek goddess of peace '*Irene*', whose themes are peace, cooperation and harmony[46]; values which suffuse and find expression in a variety of religious, philosophical and political texts across the world. For instance, the Hebrew word for peace, *Salom,* translated in the Septuagint from *Irene* also has a wide semantic range including the notions of totality or completeness, fulfillment, harmony, security and well-being. It is an umbrella term that includes much more than just absence of war or insecurity.[47] As Dr Birch, professor of the Old Testament puts it,

> In the Old Testament the opposite of shalom is not war but chaos. Thus, concern for peace must place our opposition to war alongside an equal concern for every enemy of well-being and wholeness.[48]

The implication of these cognations is that the term peace may be viewed as an antonym to chaos, thereby including good order, stability and security within its expanse. At this stage, it is important to bear in mind that all the original endeavors in creating space law since the times of Emile Laud in 1910 were focused at reining in the chaos likely in space, once the realm opened up. The convoluted version of peaceful use was but an episodic reaction to the military circumstances of the Cold War era and not an immutable gospel.

Secondly, scholastic opinion across the ages has been unanimous in that peace has always been among humanity's highest and supreme values.[49] The notion of

---

[46]Wescott (1990).

[47]New Catholic Encyclopedia (2003).

[48]Birch (1985).

[49]Lutfy and Toffolo (2018), at 393–395.

peace stands at a very high level and it serves little purpose to reduce it to a military or security notion. The UN Charter, for instance makes a clear distinction between them by stating its prime purpose is

> To maintain international peace and security, and to that end: to take effective collective measures for the prevention and removal of threats to the peace, and for the suppression of acts of aggression or other breaches of the peace, and to bring about by peaceful means, and in conformity with the principles of justice and international law, adjustment or settlement of international disputes or situations which might lead to a breach of the peace.[50]

Security in this case is a clear adjunct to peace. Also, the drift of the language indicates a desire to keep the conceptualization of peace broad based and open. Perhaps this explains why the terms, 'threat to peace', 'aggression' and 'other breaches of peace' are kept undefined. Secondly, aggression is treated as distinct from other breaches of peace as also 'situations which might lead to a breach of peace'.

The military aspect of peaceful uses is but nothing more than a subset of what peace entails. Peace can be broken by acts of military aggression and also by other reasons. For instance, a post-conflict zone may be devoid of military presence, but it is by no means peaceful; similarly a raucous mob disturbing the peace with nothing more than heightened passion and increased decibel values has nothing military about it. The concept of peaceful in day-to-day life is wide and indicates a disturbance free environment where people go about their affairs untroubled and without troubling others. The right of enjoying peace goes with the reciprocal duty of keeping one's peace. The notion of peaceful use hence needs to be perceived against the entire backdrop of issues that restrict disturbance free, untroubled and harmonious use of space. The context has changed; the scope has expanded as also the purpose and objective.

It also needs to borne in mind that the narrow conceptualizations of peace as an antonym to war and security draw on the Westphalia concept of sovereignty which does not apply in space. Sovereignty and its attendant attribute; the unfettered right of every state to make war to secure its territory are relics of classic Public International Law and the sovereignty oriented Lotus case,[51] both of which have little place in international space law. In addition to the principles which harp on space as the common heritage of all mankind, Article II of the OST prohibits claims of sovereignty in space. Even in physical terms, space is unbridled by the concept of territorial sovereignty; GEO satellites typically cover 2/3rd of the earth, LEO satellites can observe every place over earth during their orbit and constellations of navigation satellites envelope every point on earth. Thus, if the narrow conceptualizations do not apply in legal or physical terms, there is but little reason to apply its derivatives related to war and security as rigid norms.

Lastly, going by the rules of interpretation in the VCLT, the interpretation of peaceful use of space needs to expand to ensure the existing international legislation retains its dynamism and adapts to change. The interpretation of peaceful use was

---

[50] *Charter of the United Nations*, 26 June 1945, 1 UNTS XVI (in force 24 October 1945) [UN Charter] Article 1.

[51] Kalb (2018), at 3–7.

stretched in the past to include the military interests of the superpowers and there is no reason why the interpretation cannot be stretched today to include the civil interests of the world at large. The civilian interests and uses were always there, they were dormant and now with rising commercial uses, the spotlight is on them.

After all, in the past, it was the consensus mainly amongst the super powers that determined the interpretation and application of the principle of peaceful use of outer space, so there is no reason why the principle cannot be stretched to accommodate the wide variety of civil issues in space today that commonly affect most, if not all stakeholders. It would be pertinent to bear in mind that the OST has evolved beyond a "*bilateral agreement*"[52] between the two superpowers to a multilateral agreement binding 107 countries[53] across the globe that have ratified it for peaceful purposes as they interpret it.

2.4.3 Reviewing Legislative Underpinnings of the Concept of Peaceful Use of Space

Military activities in space are today fairly well regulated and the absence of conflict in space reinforces confidence in the fact that the chances of actual conflict in space are remote. Secondly, civil and military space technologies, especially in terms of space applications like communications, navigation and observation are increasingly intertwined. For instance, the US now operates a hybrid commercial and military architecture[54] with commercial satellites providing 80% of its satellite communications.[55] Similarly, non-western states are also increasingly using commercial space applications for military needs.[56] A large number of institutes, organisations and academia are already focused on the military issues in space. However, with military activities diffused, the major issues today and for the future are the absence of binding regulations with regards to space debris, congestion in useful orbits, space traffic management, on-orbit collisions, and to an extent space tourism. All of these issues affect the world at large and are critical to continued use of space. They inhibit trouble-free peaceful use of space for everybody. Looking at peaceful use in an all-encompassing manner changes the complexion of existing space treaties and expands the scope, objective and purpose of the prevailing obligations. To begin with, the preambular norms of OST that establish the purpose of the treaty recognize:

> the common interest of all mankind in the progress of the exploration and use of outer space for peaceful purposes.[57]

---

[52]Fawcett, *supra* note 41.

[53]*Status of International Agreements Relating to Activities in Outer Space as at 1 January 2018*, UN COPUOUS, A/AC.105/C.2/2018/CRP.3 (9 April 2018).

[54]Erwin (2018).

[55]Hampson (2017).

[56]Bragg (2017).

[57]OST, *supra* note 39, Preamble.

Additionally, if one looks without a military overhang at the operative clauses of the treaty like Article 3 of OST, it becomes amply clear that exploration and use of space that disturbs peace is to be scrupulously avoided. Article 3 states:

> States parties to the Treaty shall carry on activities in the exploration and use of outer space, including the Moon and other celestial bodies, in accordance with international law, including the charter of the United Nations, in the interest of maintaining international peace and security and promoting international cooperation and understanding.[58]

Bereft of military trappings, it is quite clear that space activities that disturb peaceful use by other states and non-state stakeholders are prohibited. In fact, states are duty-bound to ensure their activities do not disturb the peaceful use of space by others.[59] The duty extends both in the realms of earth and in space. For instance, the obligation to avoid harmful interference in satellite communications extends to both inter-satellite links in space and also satellite-ground station links on earth.[60] There is no reason why it should be otherwise and by this yardstick activity like scattering ashes in space or leaving spent satellites and other junk in space are clearly prohibited. Analogically, it is the same as littering debris on highways or leaving broken vehicles unattended, activities that lead to loss of enjoyment (exercise of right) of peaceful use of common property. The right in space is common to all stakeholders and by the above yardstick, no party (state or non-state) has the right to restrict or retard the enjoyment of peaceful use of space of others. The connection of peaceful use of space to the plethora of modern issues in space becomes remote when viewed through military frames of reference, devoid of the same the proximate connection is strikingly clear.

2.4.4 *UNCOPUOS: Evolving to 'Peaceful Use' in the Broader Context*

Equally clear is the fact that the UNCOPUOS, currently the only international multilateral forum dedicated to affairs related to the peaceful uses of outer space, might itself have to evolve and adapt to the changed context of what peaceful use implies. This is particularly so with regards to commercial activities in space which are rising exponentially and carry the potential to disturb the peace in space unless constructively harnessed. The need for a platform dedicated to constructively engage the private sector is imminent in view of their increased role today and in the future. Towards this end, the UNCOPUOS, going by its 2017 report, has acknowledged the growth of the space sector and endorsed increased engagement with the commercial

[58] *Ibid.* Article 3.

[59] *Ibid.* These duties are implicitly and explicitly stated in the Preamble, Articles I, III, IV, VI and IX of OST.

[60] *Constitution and Convention of the International Telecommunication Union* 22 December 1992, Vol. 1825, 1-31251 (in force 1 July 1994) [ITU Constitution] Article 45.

sector,[61] but in view of the thousands of small satellites being launched by the commercial sector and a trillion-dollar market, the question is would the above suffice to rein in the chaos likely once the satellites are in orbit and the market burgeons to unprecedented levels? Secondly, if the private sector continues to have an increased role, how can States ensure that guidelines, codes of conduct, norms are in fact relevant to these actors and more importantly will be followed by them? These and a variety of questions abound. As the character of peaceful use evolves, the international organisation dedicated to ensuring peaceful use of space may also be expected to evolve to contain the new challenges. A variety of precedencies exist and the commercial sector, in many cases, has been part and parcel of such evolution of international organizations. For example, in the contiguous realm of aviation, commerce was important right from the beginning and accordingly, the International Air Transport Association (IATA), a trade association of world airlines, which existed since 1939 provided vital inputs for the creation of ICAO and the Chicago convention.[62] On the other hand, the ITU which existed since 1865 included the private sector in its fold through Sector Members and Associates from industry when the need arose.[63] Thus, there exists but little reason why the UNCOPUOS cannot evolve to include the commercial sector within its fold.

In view of the large numbers and wide variety of players involved in commercial space activities, UNCOPUOS could also reach out to other international organizations like the International Institute for the Unification of Private Law (UNIDROIT). The UNIDROIT's purpose is to study needs and methods for modernizing, harmonizing and coordinating private and in particular commercial law as between States and groups of States and to formulate uniform law instruments, principles and rules to achieve those objectives. Entering into a dialogue with such agencies would enable purposeful harnessing of the gains of commercialization, it would also serve to facilitate coordination and collaboration on the governance of space assets at the international level, taking into account potential linkages between the sustainability of the use of outer space and the 2030 Agenda for Sustainable Development.

2.5 Conclusion

The intensely competitive Geo-political environment of the past has been replaced by an equally intense environment of Geo-economics[64] where access to space resources is increasingly seen as a means to economic aggrandizement, social welfare by both state and non-state parties. The accent is more on the civil, commercial rather than military uses of space. In the present era of globalization, realist assumptions of the

---

[61] UNOOSA (2018), at iii.

[62] International Air Transport Association, "Early Days", online: *IATA* https://www.iata.org/about/Pages/history-early-days.aspx (accessed on 22 May 2017).

[63] Leonard (2018), at 288–289.

[64] Scholvin and Wigell (2018), at 4.

Cold-war era of geo-economics being a tool of economic statecraft is increasingly giving way to geo-economics being a means of economic integration and cooperation for both state and non-state parties. This is particularly so in view of the increased diffusion in provision of space capabilities like launch services, SATCOM which are multinational as also multi conglomerate with a wide variety of state, commercial and private players. The global space economy, currently valued at about $350 billion is expected to grow to a trillion dollars in the next few decades.[65] The incentives for civil cooperation can hence be expected to rise as opposed to military competition. At the same time, commercial cooperation and competition may be expected to rise and this aspect now demands greater regulatory focus.

Apart from this, space capabilities are no longer confined to the superpowers or only states but are used daily by people across the world for a variety of uses ranging from banking to geo-location. They serve societal needs in many ways. Thus, to restrict peaceful use of space to a military context or only among states would amount to sub-optimal application of this important principle of law. The duty of ensuring peaceful use rests with states. Enforcement (powers of sanction) would also continue to rest with states. However, the moral obligation to abide by peaceful use of space would extend to non-state actors as well. The spirit of peaceful use of space as understood in an ordinary meaning is far more important today than ever before. Interpreting peaceful use as is ordinarily understood enables far more comprehensive application of existing regulations in addressing

(a) Common vulnerabilities like environmental degradation due to the surge in small satellites and consequent debris.
(b) In harnessing opportunities related to international cooperation for the common benefit of all mankind.
(c) In constructively harnessing commercial competition for the benefit of all mankind.

The notion of peaceful use, if liberated of its military fetters, enables optimal utilization of space capabilities for all mankind. It would hence be worthwhile to view peaceful use in a more contemporary and wider context.

There has been a surge in general cooperation amongst various stakeholders in space. At the same time, the environment is quite fluid and complex with many more players and hence unanimity as obtained by UNCOPUOS in the 1960s for treaty formation can no longer be expected. On the other hand, the existing international space legislation is purposeful and resilient having outlived its original state parties like the USSR. It would be worthwhile to examine whether the existing regime, when viewed through the broader frames of reference of peaceful use is flexible enough to accommodate the challenges of modernity and provide feasible solutions. The following chapters aim to do that.

[65] Foust (2018).

References

Almond HH (1991) Emerging law of outer space: the analogy of maritime salvage. J Space L 19(2):67

Alves PG (1991) Prevention of an arms race in outer space: a guide to the discussions in the conference on disarmament, UNIDIR/91/79. UNDIR, New York. http://www.unidir.org/files/publications/pdfs/prevention-of-an-arms-race-in-outer-space-a-guide-to-the-discussions-in-the-cd-en-451.pdf

Birch BC (1985) Old testament foundations for peacemaking in the nuclear era. Christ Century 1115

Bragg B (2017) Use of the commercial space industry for military purposes by non-western states. NSI ViTTA report, pp 2–4

Christol CQ (1991a) Space law: past, present and future. Kluwer, Boston,

Christol CQ (1991b) Space law: past, present and future. Kluwer, Boston; and Report of the legal subcommittee on the work of its 5th session (12 July-4 August and 12–16 September 1966), COPUOS, A/AC. 105/35, Annex IV. Retrieved from www.unoosa.org/oosa/oosadoc/data/documents/2017/aac.105/aac.1051122_0.html

Dickson P (2001) Sputnik: shock of the century. Walker and Coy, New York

Doyle SE (2002a) Origins of international space law and the international institute of space law of the international astronautical federation. Univelt, San Diego

Doyle SE (2002b) Origins of international space law. Univelt, San Diego

Englehart AB (2008) Common ground in the sky: extending the 1967 outer space treaty. Pac Rim Law Pol'y J 17(1) 133

Erwin S (2018) Space communications: as the industry launches new products, military can't decide what to buy. Space News. Retrieved from https://spacenews.com/space-communications-as-the-industry-launches-new-products-military-cant-decide-what-to-buy/

Fawcett JES (1968) International law and the uses of outer space. Manchester University Press, Manchester

Foont BE (2007) Shooting down civilian aircraft: is there an international law. J Air Law Commer 72(4)

Foust J (2017), Is it time to update the outer space treaty?. Space Rev. Retrieved from http://www.thespacereview.com/article/3256/1

Foust J (2018) A trillion dollar space industry will require new markets. Space News. Retrieved from https://spacenews.com/a-trillion-dollar-space-industry-will-require-new-markets/. Accessed on 6 July 2018

Grego L (2012) A history of anti-satellite programs. Union of Concerned Scientists, p 2

Grotius H (2004) The free seas. Liberty Fund, Indianapolis

Hampson J (2017), The future of space commercialization. Niskanen Center Research Paper

Jakhu RS (2005) Legal issues relating to the global public interest in outer space (Unpublished). Retrieved from https://drum.lib.umd.edu/bitstream/handle/1903/7916/jakhu.pdf;sequence=1

Jasentuliyana N (2010) Keynote address on space law. In: Paper delivered at the 61st international astronautical congress, 27–28 Sept 2010. IAC-10/E7/1/1/x6237

Jenks CW (1965) Space law. Praeger, New York

Kalb R (2018) International law on the maintenance of peace: Jus Contra Bellum. Edward Elgar, Cheltenham

Kendall D (2018) Reflections on the safety, security and sustainability of outer space: the united nations. In: Canadian air and space institute ASTRO-18 conference, 15–17 May 2018

Korovin E (1934) La Conquête de la Stratosphere et le Droit International. 41 Rev. Gen de Droit International Public 675

Laude E (2002) Questions pratiques (1910) 1 Revue Juridique Internationale de Locomotion Arienne 16

Leonard B (1995) Basic facts about the united nations. United Nation Publications, New York

Lutfy MW, Toffolo C (2018) Handbook of research on promoting peace through practice, academia and the arts. IGI Global, Hershey, PA

Mandl V (1932) Das Weltraum-Recht: Ein Problem der Raumfahrt (The law of outer space: a problem of space flight). In: Kopal V (ed) Vladmir Mandl—founder of space law (1968) 11. Proceedings if IISL, Mannheim, Berlin, p 357

McDougal MS, Lasswell HS, Vlasic IA (1963) Law and public order in space. Yale University Press, New Haven

New Catholic Encyclopedia (2003) Peace in the Bible. Retrieved from https://www.encyclopedia.com/religion/encyclopedias-almanacs…and…/peace-bible. Accessed on 12 Sept 2018

Peebles C (1983) Battle for space. Beaufort Books, New York

Reed F (1987) The day the Rocket Died. Air Space Smithson 2(4):52

Reijnen BCM (1992) The united nations space treaties analysed. Editions Frontieres, Paris

Scholvin S, Wigell M (2018), Geo-economics as a concept of practice in international relations: surveying the state of the art. Finnish Institute of International Affairs Working Paper No. 102

Taft E (2017) Outer space: the final frontier or final battlefield. Duke Law Tech Rev 15(1)

UNOOSA (2018) 2017 Annual report. United Nations, New York

US Central Intelligence Agency (1962) Definition of peaceful uses of outer space (contingency), 13 Mar 1962. UNCOPUOS, Washington DC. Retrieved from https://www.cia.gov/library/readingroom/docs/CIA-RDP66R00638R000100150079-1.pdf

Wescott RW (1990) Reflections on the etymology of some words for 'Peace'. Int J World Peace 7(3):94

Chapter 3
Converging Interests and International Cooperation: Solutions in International Law

Abstract The number of space actors have now evolved beyond two state contestants to numerous state and non-state actors. The activities in space have also multiplied manifold. Consequently, there is an enormous diffusion of space activities, actors and the avenues for cooperation have risen as never before. This chapter seeks to examine the aspects of how cooperation in space evolved, how international law enabled the evolution, what mechanisms of cooperation exist, the extent to which these mechanisms encourage or inhibit the spurt in civil cooperation and commerce, whether existing legislation can be gainfully adapted to the changed situation, which areas are readily adaptable and which require modification or change in entirety. Legal review and reform for military activity is stagnant because of legacy issues, but the emerging areas of civil convergence, cooperation are novel and hence may be amenable to legal review and reform.

3.1 Role of International Law in Shaping the Contours of Space Cooperation

The pace of change in outer space is shockingly inconsistent. For decades, barely a few countries operated a handful of satellites, and suddenly the numbers shot up in the millennium. As of 2017, over sixty countries operate more than 1400 satellites, there are 70 government space agencies, and an increasing number of private actors.[1] The year 2017 perhaps sets the pace for the future considering that in this year, the United Nations Office of Outer Space Affairs (UNOOSA) recorded 357 objects launched into space. This is almost 50% more than ever previously recorded in a single year, and the pace can only be expected to rise in view of the surge in small satellites bringing in its wake challenges as also opportunities. As the numbers rise, the scope of space activities also widen, and traditional areas of collaboration and cooperation give way to new areas. The areas for exchange of technologies, sharing of cost, resources may be expected to expand proportionately and would be different from the past.

[1] Del Rio Vera (2017).

K. K. Nair, *Small Satellites and Sustainable Development - Solutions in International Space Law*, SpringerBriefs in Law, https://doi.org/10.1007/978-3-030-18620-3_3

The future in space today is distinctly different from what was envisaged when space law was being drafted in the 1960s. The concerns then were military, today they are commercial. For example, in 1965, Bloomfield reflects the concerns then prevalent by stating:

> A possible future significance of space, which we overlook at our potential peril, is as a new place d'armes, a sort of strategic suburb of earth to which the Cold War has already expanded and from which, as visualized by at least some military strategists, a hot war could be fought. Both superpowers have agreed in a 1963 UN resolution to abstain from placing nuclear weapons in orbit around the earth. But both are also reportedly at work developing, among other things, manned space platforms, a fundamental first step in establishing a controllable and potentially versatile military presence in space.[2]

The future of manned space platforms in those times was envisaged for military uses, the same platforms are today envisaged for commercial use. The world is changing and as succinctly noted by David Kendall, Chairman of UNCOPUOS in May 2018:

> In general, global budgets for space activities are increasing, in many cases dramatically; more and more players are entering the arena; governments are reassessing their roles and mandates with respect to how space activities are funded, managed and conducted in their countries; and the private sector has become the dominant actor in the development of new technologies and applications. The space business has changed dramatically over the past half century and the pace, as noted, is accelerating; witness the successful development and launch of the world's most powerful rocket-the Falcon heavy rocket of Space X and, somewhat on the other end of the scale, a parallel event featuring a much smaller, but nevertheless noteworthy launch vehicle from New Zealand sixteen days prior; the important point being that both developments were financed solely by private capital – no government funds were required or requested.[3]

In hind sight, what is evident today is that in a changing world, the existing legal regime has adapted. It has been able to contain a 'hot war' in space. What is equally explicit is that the regime has enabled a variety of cooperation mechanisms ranging from treaties to a variety of agreements/arrangements (regional, bilateral, multilateral, binding and non-binding) to Charters, Letters of Intent and Memoranda of Understanding among others.[4] So resilient is it that manned space platforms originally envisaged for military use adapted and transformed first into a regional followed by a global civil cooperative endeavor and are now in the process of transforming into a commercial cooperative venture. What is unclear is how the existing regime would respond to this sudden surge in commercial space activity.

In view of the above, it would be in order to examine how cooperation in space evolved, how international law enabled the evolution, what mechanisms of cooperation exist, the extent to which these mechanisms encourage or inhibit the spurt in civil cooperation and commerce, whether existing legislation can be gainfully adapted to

[2] Bloomfield (1965), 603 at 604.

[3] Kendall (2018).

[4] For details on types of cooperation mechanisms, see *Categorization of International Mechanisms for Cooperation in Peaceful Exploration and Use of Outer Space*, UNCOPUOS, A/AC.105/C.2/2015/CRP 15, (2015) at 9.

the changed situation, which areas are readily adaptable and which require modification or change in entirety. Legal review and reform for military activity is stagnant because of legacy issues, but the emerging areas of civil convergence, cooperation are novel and hence may be amenable to legal review and reform.

3.1.1 Evolution of International Cooperation in Space

Military competition among the superpowers set the grounds for international cooperation in space. The launch of Sputnik in 1957 not only heralded the space age but also the possibility of being targeted and observed from space. Though it promised much since it was launched as part of an international cooperative endeavor, the International Geo-physical Year (IGY), its launch on the SS-6 ICBM caused much consternation and concern in the US. A military overhang to activities in space was inherent both on account of the prevailing climate of a cold war as also legacy issues related to World-War-II. Consequently, it comes as little surprise that both superpowers invested vast amount of resources in creating assets to promote their respective positions. The competition even extended beyond military to demonstration of scientific prowess, so that when the US sought a jointly planned Moon shot, the Soviets reportedly did not even respond to the initiative.[5] The frosty atmosphere then prevailing is best captured by the remarks of the US to the UN:

> Unhappily this astounding progress in space science has not been matched by comparable progress in international cooperation. In the race of history, social invention continues to lag behind scientific invention.[6]

Measures for international cooperation outside the UN yielded nothing and even within the UN were met with resistance. The UN, on its part, despite the great power struggle amongst the superpowers and the frosty environment, attempted promotion of international cooperation by a variety of means including introducing new resolutions and regulations on the subject. Towards this end, the thirteenth General Assembly on 13 Dec, 1958 established an eighteen-member Ad Hoc Committee on the Peaceful Uses of Outer Space. Unsurprisingly, the Committee could do little with the Soviets boycotting and others like India and the United Arab Republic declining to take part as an expression of non-alignment. Thus, while no legal code came about, the final report adopted on 25 June, 1959, nonetheless represented the first concerted intergovernmental attempt at international cooperation in space.[7]

The fourteenth General Assembly that followed rechristened the committee to the Committee on the Peaceful Uses of Outer Space. This institutionalization of efforts within the UN yielded results and by September 1961, the Assembly passed

[5]Bloomfield, *supra* note 2 at 604–605.

[6]*Statement in the First (Political and Security) Committee* (United States Delegation to the General Assembly, Press Release 3875, 4 December, 1961).

[7]*Report of the* Ad Hoc *Committee of the Peaceful Uses of Outer Space,* UN Doc A.4141/25 (1959) at 42.

a resolution on International Cooperation in space establishing the basic principles of international conduct in space and recommended to states that the principles of international law, including the UN Charter apply to space.[8] The following year opened up to greater cooperation with a dialogue between the leaders of the two superpowers on some non-contentious areas like weather satellites, space medicine and manned space exploration. Apart from facilitating dialogue, institutionalization served the purpose of bringing conflicting opinion to the table with both superpowers opening dialogue on the question of peaceful use of space and space disarmament at the legal subcommittee. By 1963, one is witness to a considerable narrowing of the central differences and the US and Soviet Union undertook joint space ventures in weather satellites, joint experiments with communications (*Echo-II*) and the World Magnetic Survey. Another milestone in the same period was the GA resolution of 17 October, 1963, welcoming expressions by the US and Soviet Union of their intention to not station in space any objects carrying nuclear weapons or weapons of mass destruction and calling on all states to refrain from stationing such weapons in space.[9] This expression of intention by the superpowers was to initially find reflection in the Partial Test Ban Treaty of 1963 and later in Article IV of the Outer Space Treaty-1967. At this stage, it would be pertinent to note that the question of peaceful use of space and the expression of disarmament were issues in the rubric of international cooperation that besotted only the super powers and not other nations. The super powers shaped the contours of space law and ensured its growth in a direction aligned to their interests. International cooperation, however, is driven by a variety of factors and despite the lack of a legal mandate, nations perceived it as desirable and pursued it on their own terms. For instance, India, despite its strategic alliance with the Soviets cooperated with the US and France to establish the Thumba Equatorial Launch Station in 1963.[10] Thus, the period from the late 1950s to 1960s witnessed episodic bouts of cautious cooperation between the superpowers and this continued until the Outer Space Treaty of 1967.

3.1.2 Cold War Space Treaties; Détente to International Cooperation

The reason for development of new treaty law in space were manifold; it was meant to deal with the rapid technological advances as also stem the fear of war and promote international cooperation. The response was timely and universal. During this period, one is witness to elements of conflict, competition, and cooperation in the relations

[8] *International Cooperation in the Peaceful Uses of Outer Space* UNGA Res 1721 (XVI), 16th Session (20 December 1961).

[9] *Question of General and Complete Disarmament*, UNGA Res 1884 (XVIII), 18th Session (17 October 1963).

[10] Siddiqi (2015).

between the two superpowers. Contextualizing in terms of game theory, Jonathan Galloway poses the question:

> Were the two superpowers in a zero-sum conflict in which one side would win and the other lose, or were they in a non-zero sum game in which, through learning, cooperation and peaceful competition might evolve?[11]

In hindsight, it is evident that the Cold War turned out to be a non-zero sum game leading to more benefits, peaceful uses, and spinoffs for the United States, Russia, and all of mankind. The 1967 Outer Space Treaty, negotiated under the United Nations auspices, prohibited sovereignty in outer space, thus removing one of the classic causes of war-the search for new territory. It mandated that the exploration and use of outer space would be for the benefit and in the interests of all countries and be the province of all mankind.[12] Further, it was an arms control treaty as well, for it banned State parties from placing weapons of mass destruction ("WMDs") in orbit.[13] For these and other reasons, the Treaty was a revolutionary advance in the law and reflected the spirit of detente, which existed at intervals during the Cold War period.

The Outer Space Treaty also had articles which pointed the way forward to the four other United Nations/COPUOS negotiated treaties-the Rescue and Return Agreement of 1968,[14] the Liability Convention of 1972,[15] the Registration Convention of 1976,[16] and the Moon Agreement of 1979.[17] National legislation and subsequent UN Resolutions all derive from the initial burst of energy and purpose in this formative period with the exception of early ventures by the space powers such as NASA legislation in the United States in 1958 and the formation of Comsat in 1962 and Intelsat in 1964.[18] National legislation followed much later in other countries ranging from Australia to the United Kingdom. In general, the era of treaty formation for the law of outer space appears to have ended with the Moon Agreement of 1979 and has been replaced by more specific and incremental steps including memoranda of understandings, framework agreements, voluntary regimes, codes of conduct, and case law decisions. At levels lesser than global treaties, regional conventions aimed more at sharing of resources and cost rather a spirit of *Detente* such as the Convention

---

[11]Galloway (2008), 516 at 517.

[12]*Treaty on Principles Governing the Activities of States in the Exploration and Use of Outer Space, including the Moon and Other Celestial Bodies*, 27 January 1967, 610 UNTS 205 [OST], Article I.

[13]*Ibid.* Article VI.

[14]*Agreement on the Rescue of Astronauts and the Return of Objects Launched in Outer Space*, 22 April 1968, 672 UNTS 119 [ARRA].

[15]*Convention on International Liability for Damage Caused by Space Objects*, 29 March 1972, 961 UNTS 187 [LC].

[16]*Convention on Registration of Objects Launched into Outer Space*, 14 January 1975, 1023 UNTS 15 (in force 15 September 1956) [Reg Conv].

[17]*Agreement Governing the Activities of States on the Moon and Other Celestial Bodies,* UN Doc. A/34/664, New York, 5 December 1979 [Moon Agreement].

[18]*National Aeronautics and Space Act of 1958,* 42 U.S.C. ch. 26 § 2451 et seq. (1958); *Communications Satellite Act of 1962*, 47 U.S.C. ch. 6 § 701 et seq. (1962); and *International Telecommunication Satellite Organizatoin (INTELSAT) Privileges and Immunities Order,* SOR/81-907.

of the European Space Agency in 1975,[19] Arabsat in 1976,[20] and EUMETSAT in 1983[21] also came about reflecting the desire for cooperation.

3.1.3 Evolving Contours; Peaceful Use to Commercial Use

The space treaties served their purpose eminently enabling cooperation among states in many ways and endured over half a century of presence and application. However, their prime areas of focus were state activities and the need to inhibit state competition from spiraling onto space. For instance, even in 1994, the then UN Secretary General, Boutros Boutros-Ghali while speaking on international space cooperation spoke primarily on security and arms control issues reflecting a continued emphasis on the military rather than civilian or commercial areas of cooperation.[22] By contrast, the present UN Secretary General, António Guterres, during the June 2018, UNISPACE +50 Conference made no mention of the same and instead emphasized a desire to "*harness this spirit of cooperation to use space science and technology to advance sustainable development and build a better world for all.*"[23]

The change is apparent, the Cold War is past and the present is witness to a variety of space players and intense commercial competition. In keeping with the changed context, even though peaceful use continues to be a cardinal principle in the framework of space law, the UNCOPUOS mandate has expanded to topics that reflect many more regulatory priorities. Consequently, the objective of states with regards to peaceful uses has expanded to address commercial activities and conduct as well.[24]

Within the space treaties, the sole provision dealing with the issue is Article VI, which connects the activities of non-state actors to the state by imposing international responsibility on the state for any acts, misconduct by the former. This provision, on one hand forms the basis for state responsibility and introduction of national legislation to fulfill international obligations. However, the provision does not address the vast panoply of challenges arising out of the growing participation of private space industry. The number of private players and activities in space outstrip those of the state and even the most symbolic bastion of state enterprise, the International Space Station is not immune to the change, given the recent calls for its commercialization. The evolution is dramatic considering that the ISS, once built for military purpose

[19] *Convention for the Establishment of European Space Agency and ESA Council*, Ref CSE/CS (73) 19, rev 7 (in force 30 October 1980).

[20] *Agreement of the Arab Corporation for Space Communication*, (in force 16 July 1976) [ARABSAT] in Telecommunications Journal (IX/1977) at 422.

[21] *Convention for the Establishment of a European Organisation for the Exploitation of Meteorological Satellites* [EUMETSAT] (in force 19 June 1986), online: https://www.eumetsat.int/website/wcm/idc/idcplg?IdcService=GET_FILE, (accessed on 12 October 2017).

[22] Ghali (1994), 265.

[23] Guterres (2018).

[24] Bourley (1988), 131.

transformed to a multi-national civil enterprise and now is in the process of being converted into a commercial enterprise. Any discussion on international cooperation would be incomplete without a perusal of the legal framework holding this enterprise in place and the same is attempted below.

3.1.4 *The International Space Station: From State to Commercial Cooperation*

The original conceptualization of space stations was for military use. In 1952, Von Braun proposed the space station concept as a platform that would undertake reconnaissance of threat countries and also be equipped with missiles for attack or defence. Towards the 1960s, these projects took the form of intelligence gathering platforms like the Manned Orbiting Laboratory (MOL) for the US and Almaz for the Soviets.[25] Assembly of the International Space Station (ISS) began with the launches of the Russian control module Zarya on November 20, 1998, and the U.S.-built Unity connecting node the following month, which were linked in orbit by U.S. space shuttle astronauts. In mid-2000 the Russian-built module Zvezda, a habitat and control centre, was added, and on November 2 of that year the ISS received its first resident crew, comprising of Russian and US astronauts. It has been continuously habited since then.

The ISS includes contributions from 15 nations. NASA (United States), *Roscosmos* (Russia) and the European Space Agency are the major partners of the space station who contribute most of the funding; the other partners are the Japanese Aerospace Exploration Agency and the Canadian Space Agency.[26] What keeps the ISS together is a 1988 Inter Governmental Agreement (IGA) signed by 15 states.[27] Additionally, various Memoranda of Understanding (MoU) among the partners deal with the numerous details of the supply of ISS parts and its construction. This results in a complex web of legal and sub-legal arrangements the prime aspects of which are described below.

Firstly, a three-tier legal framework governs the ISS which consists of[28]:

- One Intergovernmental Agreement (IGA) signed 29 January 1998 which stipulates that UN Conventions governing Space activities apply to ISS Cooperation.

---

[25]Chaldek (2017), at 2, 9–10.

[26]National Aeronautics and Space Administration, "International Cooperation" online: https://www.nasa.gov/mission_pages/station/cooperation/index.html, (accessed on 10 November 2017).

[27]*Agreement among the United States of America, Governments of Member States of the European Space Agency, the Government of Japan, and the Government of Canada on Cooperation in the Detailed Design, Development, Operation, and Utilization of the Permanently Manned Civil Space Station*, 29 September, 1988, [IGA].

[28]For details, see US Department of State (1998).

- Four Memoranda of Understanding (MoU) between NASA and each of the four cooperating space agencies of Russia, Canada, Europe (signed 29 January 1998) and Japan (signed 24 February 1998).
- Various Implementing Arrangements (IAs) concluded as and when the need arises between NASA and another Cooperating Agency.

Further, Article 1 (Paragraph1) of the IGA lays down that the objective of the IGA is:

> to establish a long-term international cooperative framework among the Partners, on the basis of genuine partnership, for the detailed design, development, operation, and utilization of a permanently inhabited civil international Space Station for peaceful purposes, in accordance with international law. This civil international Space Station will enhance the scientific, technological, and commercial use of outer space.

Secondly, Article 1 (Paragraph 2) states that "*the Partners will join their efforts, under the lead role of the United States for overall management and coordination, to create an integrated international Space Station*".[29]

Evidently, the objectives of the ISS include enhancing commercial use and while it may be an integrated endeavor, the lead role clearly rests with the US. Apart from cooperation and partnerships for creation, operation and utilization of the ISS, the IGA also pays considerable emphasis on the evolutionary character of the ISS in Article 1 (Paragraph 4) by stating that

> The Space Station is conceived as having an evolutionary character. The Partner States' rights and obligations regarding evolution shall be subject to specific provisions in accordance with Article 14.[30]

Article 14 further lays down specific rights and obligations in respect of partner states including the need for partners to cooperate in proposals for additions of evolutionary capabilities. The ISS has evolved in scientific and technological terms, there is but little reason to not evolve in commercial terms.

The net effect of this is that for evolution, operation, utilization, as also for peaceful purposes, in accordance with international law to enhance the scientific, technological and commercial use of outer space, the US, as the lead player, can sell the ISS. In a recent development, press reports in February 2018, announced US plans to stop funding for the International Space Station (ISS) after 2024 and turn the station over to the private sector.[31] Further reports in June 2018 indicated that the NASA administrator was in talks with various international companies for commercial management of the ISS and many large companies were interested in getting involved through a consortium.[32] The intent for sale is not known to be anything apart from continued scientific, technological and commercial use, which in effect is peaceful use. Secondly, privatization of the ISS is enhancing commercial use of space and

[29] *Ibid*. US, Dept of State, "Treaties and other International Acts 12927", Article 1.

[30] *Ibid*. Article 4.

[31] Davenport (2018a).

[32] Davenport (2018b); and Georgion (2018).

is aligned to the 'new space' line of thinking that celebrates commercialization of space. Also, because the US has the lead role for overall management and coordination, it can in consultation with partner states, take the decision and coordinate the sale. And finally, in keeping with the evolutionary character of ISS, if it changes from a military to an inter-governmental to a commercial character, that is but part of its evolution for survival.

With regards to concerns about privatization leading to loss of state control, responsibility or wanton commercial greed taking over and playing havoc with the pristine environment of space, it needs to be borne in mind that the safeguards originally envisaged and put in place by Article 2 of the IGA continue to be in place. They continue to be as effective as ever and nothing dilutes or does away with state responsibility for acts by commercial, private players. The state ultimately continues to bear responsibility and state control is just not affected. Article 2 of the IGA lays down the international rights and obligations with Paragraph 2 stating that the ISS shall be developed, operated and utilized in accordance with international law, including the Outer Space Treaty (OST), the Rescue Agreement, the Liability Convention, and the Registration Convention. In effect, privatization does not change the state of affairs with respect to state control, jurisdiction, responsibility and liability. The legal obligations, implicit and explicit in these international treaties prevail. Article 2, para 2 further specifies that nothing in the IGA modifies the rights and duties of partners in respect of space treaties, consequently, commercial gain cannot override the duty of international cooperation explicit in international law and the space treaties. Conversely, if a case for commercial international cooperation among partner states is made, as is the case here, it only promotes the principle of international cooperation further. At the same time, under Article VI, Outer Space Treaty, states shall bear international responsibility for their national activities in space and hence while asserting leadership role, the US would need to consider the legal prudence of taking over responsibility for other nation's activities in space. Similarly, as per Liability Convention, the State is responsible for compensation in case of damage by space objects and hence whether the US would like to accept liability for actions of others needs to be thought out.

The aspect of transfer of ownership in day-to-day life is inherently tricky and is no different in case of the ISS. The transfer of ownership of ISS elements/equipment does not affect rights and obligations under the IGA.[33] However, any transfer requires prior notification and requires consent of all parties of IGA.[34] Thus, the decision would require intense consultations amongst partners and can only be consensus driven. Apart from hardware, human relations are covered by Article 22, IGA whereby states retain jurisdiction over their nationals. In case of misconduct, criminal activity, the state of the alleged perpetrator is to consult the other party at its request as to their respective interests in a prosecution. The affected partner state may exercise criminal jurisdiction ninety days after the consultation (or other agreed period) if the state of the perpetrator so agrees, or fails to prosecute under its legal system. Extradition is

---

[33]IGA, *supra* note 27, Article 6.3.

[34]*Ibid.* Article 6.4.

also possible. An ISS code of conduct also applies to crew members. However, all of the above applies in case of crew of ISS member states. How this complex legal issue can be handled in case of privatization demands specific treatment. For instance, the question arises as to jurisdiction over 'space tourists' who may not be nationals of an ISS member state. As of now, the law is unclear. What can only apply is a moral obligation to refrain from acts that disturb peace which is imposed by a broad interpretation of the concept of peaceful use of space. Liability aspects are equally complex in that Article 16 of the IGA deals with cross-waivers of liability or affects the rights or obligations of partner states in exploration and use of space. However, this draws on the 1972 Liability Convention which was designed for states. Converting this complex web of liabilities from state to private, commercial character is fraught with difficulties in legal, sub-legal arrangements at international, national and private levels. The new areas of commerce in space may be expected to include space tourism, transportation of people and cargo to ISS, advertising and commercial experiments. Here also, the ISS cross-waiver would not apply to tourists and such participants who may not be from ISS member states and may have diverse backgrounds. New regulations consistent with the ISS regulations as also the Liability 'Convention would need to be formulated to cater for such contingencies. The issue is difficult but not impossible to handle. In summation, the legislative framework of the IGA and four MOUs has been flexible enough to regulate affairs well amongst the partners till date without the need for an amendment, while the Implementing Arrangements and program instruments have evolved as required.

Having dwelt upon the manner in which law has shaped the contours of international cooperation in response to changing needs, it would be in order to examine the mechanisms that enable such adaptation. The same is undertaken as below.

3.2 The Legal Framework of Space Cooperation

The law related to outer space is not just confined to the space treaties but includes aspects of General International Law, customary law, UN Resolutions, regional, bilateral and multilateral agreements. In fact, international space law draws on General International Law and hence it comes as no surprise that the UN Charter[35] and the subsequent 1970 Declaration[36] form the legal basis for cooperation in outer space. This finds reflection in no small measure in Article III OST which evokes the UN Charter stating:

[35] *Charter of the United Nations*, 26 June 1945, 1 UNTS XVI (in force 24 October 1945) [UN Charter].

[36] *Declaration on Principles of International Law Concerning Friendly Relations and Co-operation among States in Accordance with the Charter of the United Nations*, 24 October 1970, A/RES/25/2625.

> States Parties to the Treaty shall carry on activities in the exploration and use of outer space, including the Moon and other celestial bodies, in accordance with international law, including the Charter of the United Nations.[37]

By evoking the UN Charter, Articles 1, 2, 11, 13 and 56 of the Charter containing provisions regarding the obligation of member states to cooperate automatically become applicable.

3.2.1 The UN Charter and Its Scope of Application to Space Cooperation

The above is the broader view. However, some scholars argue that the Charter does not per se prescribe cooperation in any area other than in the maintenance of international peace and security.[38] Here, it needs to be borne in mind that as per the Vienna Convention treaties have to be read as a whole and with regard to its object and purpose.[39] The preambular declarations make it clear that the ends of the UN Charter include not only saving succeeding generations from the scourge of war but also ‘to promote social progress and better standards of life in larger freedom’. Freedom here is indicative of not just from military insecurities, but also other insecurities like food, poverty and other social issues and hence is qualified by the term larger. Towards accomplishment of these larger aims, the Charter seeks to employ international machinery for the promotion of the economic and social advancement of all peoples. Moreover, the Charter has institutionalized a system where states are not only required to work together for maintenance of peace and security but also coordinate their actions to develop friendly relations, achieve international cooperation in solving international problems of an economic, social, cultural, or humanitarian character and to be a center for harmonizing the actions of nations.[40] Thus, while cooperation is not included as an explicit, mandatory rule for every UN member, it is certainly a desirable standard of conduct for every member. The Charter, after all is meant to be broad in its scope, objective and purpose and towards this end it provides the generic concepts for conduct and interaction. It may be true that instances of broad and abstract language in the Charter make identification of specific duties and obligations difficult, but, even such clauses are not entirely devoid of legal significance; at least, they imply a negative duty on the parties to not disregard cooperation.

Another line of thought suggests that the principle of cooperation may entail either an ‘obligation of result’ or an ‘obligation of effort’, sometimes also dubbed

---

[37] *Treaty on Principles Governing the Activities of States in the Exploration and Use of Outer Space, including the Moon and Other Celestial Bodies*, 10 October 1967, 610 UNTS 205, TIAS. 6347, 18 UST 2410, Article III.

[38] Voronina (2016).

[39] *Vienna Convention on the Law of Treaties*, 23 May 1969, 1155 UNTS 331, 8 ILM 679 (in force 27 January 1980), Article 31(1).

[40] UN Charter, *supra* note 35, Article 1, para 1 to 4.

'persuasive obligation.[41] The former presupposes that a hypothetical duty to cooperate would only be complied with if a particular result is achieved—in this case, if actual cooperation is the result of a request to do so—and is usually provided for in legally binding documents. The latter, by contrast, means that a duty to cooperate merely requires States to be willing to consider cooperation in good faith when a request is made, therefore, without any specific obligation to enter into actual cooperation with the requesting State. Against this background, the United Nations Charter does lay down the principle of cooperation but does not establish a general obligation to cooperate. At the same time, it does establish an 'obligation of effort' by way of inclusion of cooperation as one of the Organization's purposes,[42] by way of entitling the General Assembly to take relevant steps in promotion of cooperation,[43] and by way of creation of the Economic and Social Council charged with creation of conditions of stability and well-being, which are necessary for peaceful and friendly relations.[44] While none of the relevant articles create a clear-cut obligation to cooperate, or more specifically the 'obligation of result', they all signify the importance of cooperation in international relations and introduce the principle of cooperation 'obligation of effort' as an indispensable part of the modern international legal order. The 'obligation of result', therefore, is limited to States' obligations to cooperate in the maintenance of international peace and security, whereas cooperation in all other areas, while being considered beneficial, is left to the discretion of States, subjecting them solely to the obligation to consider cooperative proposals in good faith.[45]

The Charter provisions related to cooperation were further developed by the 1970 Declaration.[46] However, the Declaration was adopted in the form of a United Nations General Assembly Resolution and hence is not a legally binding document. However, not all United Nations General Assembly resolutions are the same: some carry more weight than others and Lyall and Larsen note that the 1970 Declaration is an example of the United Nations General Assembly resolution that "*contain[s] interesting and valuable statements, some of which purport to be statements of international law*."[47] The 1970 Declaration was adopted by consensus. It has been opined that consensus per se does not mean anything, because States favor this method owing to the absence of necessity to take a strong position on a particular issue, and hence it relieves them from taking on any specific obligation.[48] However, it is widely held that resolutions adopted without voting or by an overwhelming majority of States might be used to

[41] Voronina, supra, note 38, at 23.

[42] UN Charter, *supra* note 35, Article 1.

[43] *Ibid.* Articles 11, 13.

[44] *Ibid.* Articles 55, 56, 61.

[45] Voronina, *supra*, note 38, at 23.

[46] *Declaration on Principles of International Law Concerning Friendly Relations and Co-operation among States in Accordance with the Charter of the United Nations*, GA Res 2625 (XXV), UN GA 25th Session, UNDOC A/RES/25/2625 (24 October 1970).

[47] Lyall and Larsen (2009), at 44.

[48] Suy (1977), at 260.

identify the emergence of a customary norm.[49] This, in addition to the fact that the document is titled as a 'Declaration' rather than a 'Recommendation' indicate the importance of the decision. More specifically, this form is sometimes deemed appropriate for codification of existing customary international law or general principles of law.[50] The Memorandum of the United Nations Office of Legal Affairs on the Use of Terms "Declaration and Recommendation" states:

> In the United Nations practice, a 'declaration' is a formal and solemn instrument, suitable for rare occasions when principles of great and lasting importance are being enunciated, such as the Declaration on Human Rights.[51]

By comparison, a recommendation is less formal. In specific terms, the Declaration in a paragraph of the preamble mentions the principle of non-appropriation of outer space and celestial bodies, thus specifically extending application of its provisions to outer space activities. However, the Declaration is but non-binding in nature. It does not establish a general obligation to cooperate in space. It only reaffirms the 'obligation of effort' established by the UN Charter and hardly more.

3.2.2 *International Cooperation and the Space Treaties*

The broad and abstract principles of cooperation contained in the UN Charter and Declaration have been deftly interwoven into the space treaties. Not only is explicit mention made of applicability of international law and the UN Charter in Article III OST and Article II of the Moon treaty, even the breadth of the abstract notions contained in them finds expression in the treaty language. For instance, both Article III, OST and Article II of the Moon Treaty repeat the wording of the Charter stating that space activities shall be carried out in the interest of maintaining international peace and security as also in promoting international cooperation and understanding. Therefore, before going into the specific clauses on cooperation, it makes sense to assess the meaning of the broad, abstract clauses contained in the treaties. A safe starting point in construing these clauses is to state that they are an integral part of the agreement and thus necessarily share their binding nature. Secondly, as Rudolph observes, while it is true that certain broad clauses of cooperation in treaties of a highly political nature may be so abstract that it is difficult to ascertain any specific duties and obligations inherent in them, this is not the case with the space treaties.[52] The duties and obligations in the space treaties are plainly laid down. It is quite doubtful, hence, whether the treaties as a whole fall into the category of highly political treaties which need to be interpreted in a narrow manner. The breadth of the clauses for international

---

[49]Conforti and Labella (2012), at 35, 42–43.

[50]Hahn (1997), at 1334.

[51]*Memorandum by the Office of Legal Affairs: Use of the Terms Declarations and Recommendations*, Commission of Human Rights, E/CN.4/L.610, (02 Apr 1962) online: https://digitallibrary.un.org/record/757136/files/E_CN.4_L.610-EN.pdf, accessed 22 December 2017.

[52]Dolzer (1985), 527.

cooperation draws on what Jakhu calls 'the concept of intentional ambiguity'[53] and is by design rather than default. Consequently, disregarding the broad structure of the treaties in construing their abstract clauses of co-operation may be both misleading and inappropriate. It is hence necessary to take into account the way in which the treaties establish more specific forms of obligations, i.e. to view the abstract clauses of co-operation against the background of those rules which have been phrased in a more specific context. The fundamental point inherent in such an approach is that it would be against the logic inherent in a treaty's structure to construe the abstract clauses of co-operation more broadly than those individual norms of co-operation which have been included explicitly in the treaty, and which generally are covered by the subject matter addressed by the abstract co-operative clause. Individual actions which clearly frustrate the object and purpose of the Space Treaties to lay the foundation for the development of an intensified co-operation would thus be inconsistent with the spirit and the text of the treaties. Unilateral actions which would prejudice the exploration and use of space for a longer period to the detriment of the interests of the international community would therefore be in violation of the treaties. The same would apply with respect to an unqualified refusal of a State party to participate and contribute in further negotiations on the development of a space regime responsive to the interests of the international community.

Narrowing down to the specifics, it needs to be borne in mind that the OST is a treaty of principles and it should be construed as such. Articles III, IX, X, and XI declare the need to cooperate, to maintain international peace and security (Article III), to have regard for corresponding interests of other parties (Article IX), to allow observation of space objects (Article X), to disseminate information about space activities (Article XI). The overarching provision for interpretation of these articles is Article I which declares that exploration and use of outer space and celestial bodies as the 'province of all mankind'. The concept of the 'province of all mankind' is different from the 'common heritage of mankind' concept, where the latter is based on the presumption that space exploitation can take place only within the limits of specific international regime,[54] while the former is focused on providing equal access to all States by promoting equal participation in its use and exploration unless specific obligations have been agreed upon. All other Outer Space Treaty provisions are inseparable from the concept of the 'province of all mankind' and need to construed accordingly.

Going beyond the conceptual underpinnings, with regards to the strictness of binding language, the mandatory character of an obligation is set forth only in very few of these clauses. Article XIII of the Moon Treaty states without qualification that a State which learns of the crash landing or unintended landing of a space object not launched by it shall promptly inform the launching party. Such clear language, however, is conspicuously absent in other clauses concerning specific co-operation in specific areas. For example, Article X of the Outer Space Treaty which regulates observation of the flight of space objects by third parties, limits the respective obli-

[53] Jakhu (2005), at 6.

[54] von der Dunk (2014), at 58.

gation of the launching State by stating that relevant requests shall be considered on the basis of equality, and it is added that an agreement between the States concerned shall establish the conditions of such observations. Another area of co-operation to be considered in this context relates to the important issue of information sharing. Article XI obliges the States parties to inform the international community about the nature, conduct, locations and results of their activities; however, this obligation is limited by the proviso that such information must be given only "*to the greatest extent feasible and practicable*".[55] On similar lines, Article V of the Moon Treaty adopts the same language regarding information obligations, but it has added provisions for the time at which—such information shall be furnished. Evidently, the mandatory obligations, even though few and far in between, are phrased in heavily guarded language which leaves a broad margin of appreciation by member states. This makes it clear that the treaties are not designed to be proscriptive in nature; they are meant to be applied in tandem with the guiding principles.

A bare reading of the OST makes it apparent that one of the prime purpose of the treaties is to ensure global access to space resources and prevent deprivation of any state of the opportunity to explore and use space for peaceful purpose. This perhaps explains why Article I specifically includes the aspect of 'province of all mankind' in its declaration that the exploration and use of space shall be carried out for the benefit and in the interests of all countries irrespective of their degree of economic or scientific development. Further, perhaps, in deference to scientific investigation in space being a means of economic and social upliftment, it lays down a clear obligation to cooperate stating that:

> There shall be freedom of scientific investigation in outer space including the Moon and other celestial bodies, and states shall facilitate and encourage international cooperation in such investigation.[56]

The above explicit duty of cooperation gains significance when read in conjunction with the preambular declarations which narrate a variety of concerns lying behind the decision to create the treaty. These make the overall scope, objective and purpose of the treaties clear. For instance, the first five opening preambular statements make it clear that not only is there a recognition of "*the common interest of all mankind in the progress of the exploration and use of outer space for peaceful purposes*",[57] there is also a belief expressed that such cooperation would contribute to development of mutual understanding and strengthening of friendly relations. Scientific investigation is the prime enabler of optimal utilization of space resources, a stepping stone to economic and scientific development and when perceived in this context, the obligation to cooperate for harmonious advancement is clear.

---

[55]OST, *supra* note 12, Article XI.

[56]*Ibid*. Article 1.

[57]*Ibid*. Preamble.

3.2.3 The 1996 Declaration on Space Cooperation

The Declaration on International Cooperation in the Exploration and Use of Outer Space for the Benefit and in the Interest of all States, Taking into Particular Account the Needs of Developing Countries adopted as United Nations General Assembly Resolution 51/22 on 13 December 1996 is another landmark effort in the pursuit of international cooperation in space. Some eminent scholars consider it to be a "*general framework for international cooperation.*"[58] The Declaration recalls the relevant provisions of space cooperation in the UN Charter, the space treaties, resolutions and recommendations of UN conferences that have been discussed earlier. The distinguishing factors include pronouncements like:

> States are free to determine all aspects of their participation in international cooperation in the exploration and use of outer space on an equitable and mutually acceptable basis. Contractual terms in such cooperative ventures should be fair and reasonable.[59]

It confirms that international cooperation should be in full compliance with international law including the Charter of the United Nations and the Outer Space Treaty. It, however, by contrast to the first draft of the Declaration presented to the COPUOS Legal Subcommittee in 1991, abandons the approach of forcing countries into cooperation. Scholars have argued that:

> the only constructive but simple reason [for that] should have been that international cooperation should not be forced upon countries, because without shared interests cooperation cannot be fruitful.[60]

It has been further argued that "*the most important political lesson might be that international cooperation neither can nor should be forced upon States.*"[61] "International cooperation according to this Declaration is characterized by the free choice with respect to modes of cooperation and the renouncement of so-called "forced cooperation" as well as any forced transfer of technology."[62] Almost 40 years following the beginning of international space cooperation States adopted a legally non-binding, though widely supported document summarizing the principle of cooperation as applied in outer space exploration.

---

[58] Benkö and Schrogl (2000), 74.

[59] *Declaration on International Cooperation in the Exploration and Use of Outer Space for the Benefit and in the Interest of All States, Taking into Particular Account the Needs of Developing Countries*, GA res. 51/122, UN Doc. A/AC.105/572/Rev. 1 (1996).

[60] Benkö and Schrogl (1993), at 213.

[61] *Ibid.*, at 215.

[62] Benkö and Schrogl, *supra* note 58, at 74.

3.3 Examining Effects of Legal Mechanisms on International Space Cooperation

The importance of law in creating institutions and facilitating international cooperation in space is evident since the beginning of the space age. The principle of international cooperation has been clearly stipulated in various instruments including those adopted under the framework of the United Nations. For example, the General Assembly resolution which established an ad hoc Committee on the Peaceful Uses of Outer Space in 1958 requested it to report to the General Assembly on the:

> area of international co-operation and programs in the peaceful uses of outer space which could be appropriately undertaken under the UN auspices" as well as "the future organizational arrangements to facilitate international co-operation in this field.[63]

Ever since, the Committee has been encouraging States to act collectively to promote the peaceful exploration and use of outer space through a variety of mechanisms. Part of such mechanisms are found in the United Nations treaties on outer space, the sets of declarations and principles on outer space activities, General Assembly resolutions and other relevant documents relating to the peaceful exploration and use of outer space. Likewise, States and international organizations have initiated various programs through the conclusion of multilateral and bilateral agreements suitable for the specific programs concerned, which have further developed the legal basis for space cooperation.

3.3.1 Resilience of Legal Mechanisms Enabling Adaptation

Mechanisms employed by States are numerous in number and of wide variety in nature, form and substance. Some cooperative projects are conducted by a multilateral agreement or a set of agreements among States which could be either legally binding, legally non-binding, or a combination of both. There are also cases where multilateral cooperation is carried out within the framework of international intergovernmental organizations, including the United Nations and its specialized agencies, international intergovernmental organizations other than the United Nations, and other types of forums, such as regional and interregional mechanisms for cooperation. Other cases represent bilateral partnerships based on either legally binding or legally non-binding agreements. In non-legally binding instruments, there are also conditions and recommended standards for space collaboration. However, the guiding principles invariably are those contained in international law including the United Nations treaties on outer space.

---

[63] *Question of the Peaceful Use of Outer Space*, Res 1348 (XII), UNGA, 13th Session (14 November 1958).

These wide variety of mechanisms enable an equally wide scope of cooperation in many areas ranging from planetary exploration to space applications and debris mitigation among other areas. It also enables assistance to developing countries to obtain space assets including supplying satellites and launch services, constructing ground facilities and providing personnel training. Since the number of mechanisms are large, apart from a wide variety of areas, a wide and varied number of actors are also accommodated in the folds of the regulations. For instance, the United Nations often features as a platform of international cooperation and also an independent actor participating in international cooperative programs. Secondly, in addition to States and international organizations which are recognized essential actors in cooperative mechanisms, the mechanism is flexible enough to facilitate the entry and advance of commercial and private actors as well.

3.3.2 Adapting to Change: Commercial Cooperation

It could be surmised that among the prime factors enabling this flexibility and accommodation of diverse players is the expansive scope and ingenious draft of the provisions in the space treaties. For instance, the preamble recognizes that the notion of international cooperation is not just limited to states by declaring that "*cooperation will contribute to the development of mutual understanding and to the strengthening of friendly relations between states and peoples*".[64] The remarkable aspect here is a clear reference to peoples in addition to states which clearly indicates a desire on part of the drafters to keep the scope of cooperation wide enough and well beyond just the confines of the state to include space activities by private parties as well. This spirit of all-encompassing cooperation further finds resonance in Article I which states:

> The exploration and use of outer space…shall be carried out for the benefit and in the interests of all countries, irrespective of their degree of economic and scientific development, and shall be the province of all mankind…[65]

Article I also emphasizes on the 'benefit and interest' clause, though the significance and exact content of the clause is unclear. What is clear, however, is that the basis for this wish is already enshrined in the UN charter[66] which states,

> We the People of the UN…determined…to promote social progress and better standards of life in larger freedom…and for these ends…to employ international machinery for the promotion of the economic and social advancement of all peoples.[67]

Space activities bring manifold benefits to all peoples. For instance, commercial SATCOM benefits people across the world in many ways as also commercial launch

---

[64]OST, *supra* note 12, Preamble.

[65]*Ibid.* Article I (1).

[66]Brunner and Soucek (2011), at 326–327.

[67]UN Charter, *supra* note 35.

and indeed space tourism by increasing space access to many more people. It may be noted here that the benefit clause does not forbid commerce or making profit. It only seeks to ensure that the benefits should flow to all. Here again, emphasis is paid not only on sharing of the benefits among 'all countries' regardless of their level of development, but also 'all mankind' indicating that the *ratio* is to not limit sharing of benefits to states but also expand to non-state actors as well. As commercial activities in space rise, a larger number of space products and services would enter the market and commercial cooperation can be expected to rise proportionately.

The legal meaning of 'benefit' according to Black's law dictionary and various other legal sources is 'profit' or 'gain'.[68] In normative terms also, the prime interpretation is the same and hence it could be surmised that the existing regime of international space law is not inherently averse to commerce in the wider sense. The term commerce does not find specific mention in the space treaties partly since commerce was not on the anvil when the treaties were created and since the prime players then were steeped so deep into competition that the possibility of commercial cooperation was too remote to merit specific mention in the treaties. The treaties were designed for universal application. Therefore, the scope of the treaties is not just state interest but global public interest and this imposes international obligations *erga omnes* on all states to promote commercial cooperation as well. The treaties were framed to be wide enough to be responsive to change. Commercial activity is not specifically mentioned but it has been alluded to with a purpose in mind and hence it makes sense to interpret the treaty principles in light of their contextual applicability in the present rather than adhere doggedly to past inferences.

Here, it could be argued that despite the wide scope of the space treaties, most private enterprise is for commercial aggrandizement and not social welfare, consequently practical application of the spirit of cooperation may be difficult. However, it needs to be borne in mind that the idealist spirit of Article I is tempered by the pragmatic approach of Article VI that vests states with the ultimate responsibility for non-state activity. State responsibility under the aegis of Article VI is not confined to just licensing of activities to adhere to international norms but also to regulate domestic activities holistically. This normally extends to social welfare, commerce and the legislation makes it incumbent upon states to ensure that the spirit of cooperation enshrined in the space treaties is observed. In fact, domestic legislation of many nations makes it a point to extend the societal and commercial benefits to all citizens. For example, nations like India always had social welfare, commerce as a primary goal and this is evidenced in its commercial state enterprise '*Antriksh Corporation*'[69] built on the legislative foundations of Indian domestic law and the space charter to churn out commercial benefits for its citizens. On the other hand, states like Luxembourg have also interpreted the space treaties dynamically to create a thriving

---

[68] "Benefit", *Blacks Law Dictionary*, 7th ed, online: https://thelawdictionary.org/benefit/; and "Benefit", in *Merriam Webster Law Dictionary* online: https://www.merriam-webster.com/dictionary/benefit.

[69] For details, see website of Antriksh Corporation, online: https://www.isro.gov.in/about-isro/antrix-corporation-limited (accessed 30 December 2017).

commercial industry. By venturing into the realm of appropriation of space resources, Luxembourg has courted controversy on a sensitive topic. However, this does not in any way reduce the significance of the fact that Luxembourg sought to come about with domestic legislation in August 2017 primarily to regulate and enhance its benefit from space by entering into cooperative ventures with other states.[70] It also sent a signal to foreign investors that Luxembourg's regulatory framework was transparent and provided stability for potential investors. To this end, the existing space treaties provide a certain level of certainty, guidance and flexibility for purposeful adaptation to changed circumstances. Luxembourg has evidently made good use of the existing space treaties considering it has attracted over 60 start ups and attracted around $13.3 Billion in foreign investments.[71]

Nevertheless, instances like the above are not new. There has always been conflict between the exclusive interests of states and the inclusive interests of all humanity. States do frequently try, and at times succeed in advancing their exclusive interests at the cost of the wellbeing of the international community. For instance, as Jakhu points out:

> The 1961 UNGA Resolution 1721 (D) declares that communication by means of satellites should be available to the nations of the world as soon as practicable on a global and non-discriminatory basis. Giving effect to this Resolution, INTELSAT and INMARSAT organizations under their respective treaties provided international public telecommunication services of high quality and reliability on a non-discriminatory basis to all areas of the world. This provided highly valuable life line services to a large number of developing countries that did not have adequate telecommunications infrastructure. Such non-discriminatory universal access was eliminated by the privatization (actually nationalization) of both INTELSAT and INMARSAT, primarily under the 2000 US ORBIT Act which forced their dismantlement.[72]

Thus, instances like the above are exceptions and in most cases the principles of cooperation in the space treaties are adhered to. By providing an element of certainty, uniformity, and a standard of application they enable guidance and flexibility of application across the board. These legal mechanisms have enabled a variety of cooperative non-state endeavors across the world in a variety of areas. The weight of evidence indicates that states cooperate when a convergence of interest arises. For instance, going by the UN report on space cooperation (2015),[73] states tend to readily cooperate in areas of common interest like disaster management, space exploration, earth observation where the wellbeing and safety of international society as a whole is concerned. In such cases, the subject Resolution is good enough to simulate states to enter into coordination mechanisms.

---

[70] For details, see website of Government of the Grand Duchy of Luxembourg, "Space Resources.lu" online: https://spaceresources.public.lu/en.html (accessed on 03 January 2018).

[71] Aliya Ram, "US and Luxembourg Frame Laws for New Space Race" *Financial Times* (19 October 2017).

[72] Jakhu (2009), at 71–77.

[73] *Categorisation of International Mechanisms for Cooperation in Peaceful Exploration and Use of Outer Space*, UNCOPUOS, UN DOC A/AC.105/C.2/2015/CRP 15, (2015) at 9.

The only convergence of interest in the past was on space weather and coordination mechanism were limited to sharing of meteorological data, scientific studies.[74] Today, a paradigm shift is evident in a diffusion of space capabilities related to global public interest uses like satellite navigation where GPS and GIS applications serve people regardless of national boundaries as also SATCOM, Search and Rescue uses and meteorological applications.

A convergence of global public demand for space applications, services and commercial acumen in meeting the demand lead to a variety of cooperative mechanisms across the world that draw on the broad guidelines of the space treaties to conduct their affairs. The resilience of the space treaties is demonstratively apparent in their continued adherence by space actors and are evidenced in the rise of multilateral coordination mechanisms like the Group on Earth Observation (GEO),[75] International Charter on Space and Major Disasters,[76] International Space Exploration Coordination Group (ISECG)[77] and the Committee on Earth Observation Satellites (CEOS).[78]

It is also evident in the cooperation and coordination mechanisms of the African Union which in 2017 sought to rise beyond state affiliation to promotion of international cooperation to serve the entire African continent rather than just states.[79]

A similar convergence of continental interest in space earlier existed only in the European continent but is now increasingly visible in many more regional space cooperative mechanisms like the Asia-Pacific Regional Space Agency Forum (APSRAF) which is a partnership for cooperation among governmental and non-governmental bodies.[80] One is also witness to increased calls for cooperation among the Latin American countries[81] and in case of India, it has in recent times expanded its foot print as never before.

3.3.3 Commercial Cooperation: Examining the Case of India and the EU

The rising convergence of interests in space has led to unprecedented levels of cooperation. The levels of cooperation agreements arrived into by the US is well known;

---

[74] Res 1721, *supra*, note 8.

[75] See online: https://www.earthobservations.org/geo_community.php (accessed 07 January 2018).

[76] See online: http://www.un-spider.org/space-application/emergency-mechanisms/international-charter-space-and-major-disasters (accessed 07 January 2018).

[77] See ISECG, online: https://www.globalspaceexploration.org/wordpress/ (accessed 09 January 2018).

[78] See CEOS, online: https://ceos.org (accessed 09 January 2018).

[79] African Union (2017), at 23.

[80] See online: https://www.aprsaf.org/ (accessed 14 January 2018).

[81] Lopez (2016).

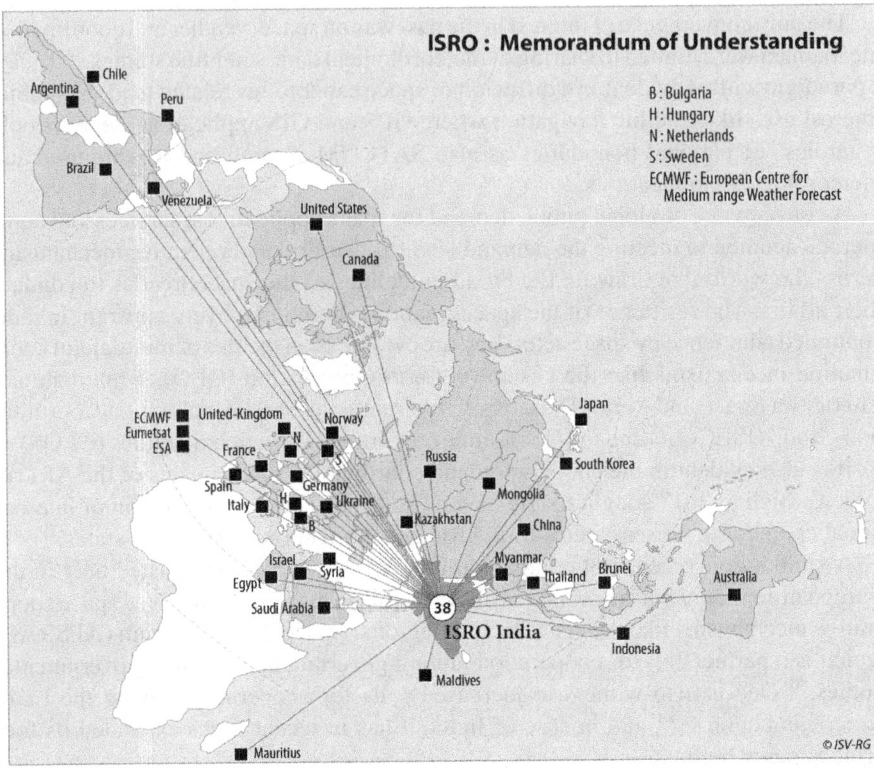

Fig. 3.1 Map of India's space cooperation

NASA has had over 3000 agreements with over 100 nations,[82] what is little known are the extraordinary levels of diversity in case of India and Europe.

India, despite not having a formal domestic law in place makes enormous use of the broad guiding principles of cooperation in international law, particularly the space treaties for pursuing its interests across the world. It's cooperative endeavors in space are largely for scientific, economic development and social welfare. There is no military cooperation at all and going by its Parliamentary report, in keeping with the spirit of using space cooperation to foster mutual understanding and strengthen friendly relations, as borne out by Fig. 3.1, it has signed bilateral agreements with 38 countries on a variety of projects ranging from the SAARC satellite to planetary exploration across the globe.[83]

[82]Badri Younes, Deputy Associate Administrator (NASA) "International Cooperation in Space: Now more than Ever" at site of NASA https://www.nasa.gov/pdf/696854main_Pres_International_Cooperation_in_Space.pdf.

[83]*Report of the Parliament of India, Standing Committee on Science, Technology, Environment and Forests*, 312th Report, Department of Space (13 March 2018).

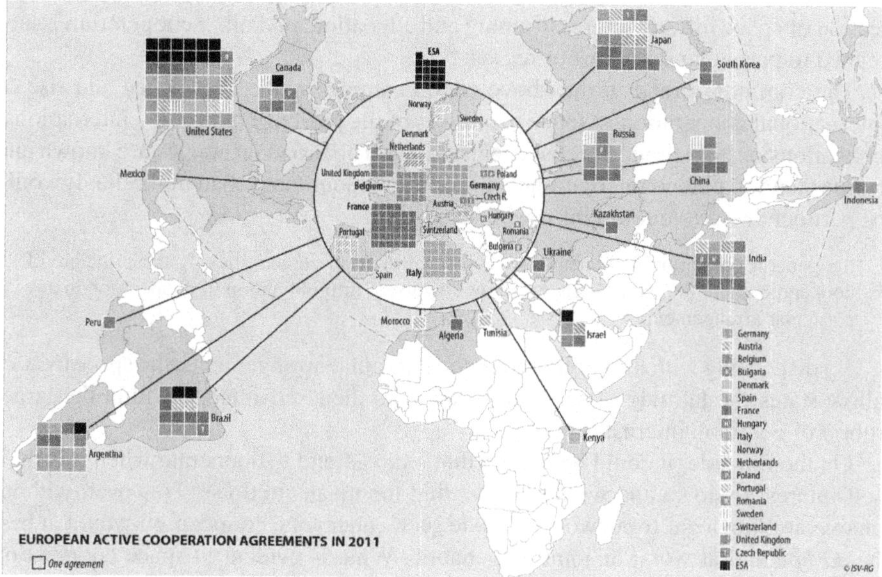

Fig. 3.2 Europe's space cooperation map

On the other hand, the European Union is a conglomeration of numerous states, some of whom have domestic legislation in place and some do not.[84] Despite this and the fact that the EU is subject to the usual stress and strains of inter-state cooperation,[85] it displays an equally expansive and diverse space cooperation map as shown in Fig. 3.2.[86] International cooperation contributes to the implementation of European space programs such as EGNOS, Galileo, and Copernicus. It also supports space research through the EU's Horizon 2020 framework program for research and development and opens up markets abroad for European space-related technology and services.

For the EU, international cooperation on space follows two complementary streams: implementing existing bilateral agreements in specific programs and pursuing non-binding cooperation related to existing programs or strategies. Regarding the former, EU space programs such as Galileo, Copernicus, or Horizon 2020 have provided a framework for international cooperation. For the latter, there is broad collaboration between the EU and non-EU countries covering areas such as space exploration through the high-level International Space Exploration Forum, the pro-

---

[84]For details, see Jakhu (2010).

[85]For a brief appraisal, see Kristin Archick, "The EU: Current Challenges and Future Prospects", *Congressional Research Service* (February 2017).

[86]Maps sourced from Isabelle Sourbes-Verger, "EU-India Cooperation on Space and Security", (2016) Instituto Afferi Internazionale, IAI Working Papers 16/38.

tection of space infrastructures, training and education, and other cooperation issues related to particular countries or regions.[87]

The common thread in the above cited examples is the prevalence and rise of international cooperation in space regardless of the presence of specific international obligations or even domestic space legislations. The broad principles are known and nations tend to pursue international cooperation under that broad umbrella. It would be pertinent here to quote Jenks, who states

> the general obligation of States to be guided in their space activities by the principle of cooperation and mutual assistance does not involve a firm obligation of cooperation in any particular arrangements.[88]

By dispensing with firm obligations and particular arrangements, the space treaties allow states the latitude of selecting what suits their self-interest within the larger rubric of common interest.

On the flip side, it could be argued that nations tend to cooperate when it is their self-interest to do so and would always find the means to do so. However, without an overarching legal framework to guide such endeavors, cooperation would at best be *ad hoc* and at worst disjoint and chaotic. What is evident in space cooperation instead is a spectrum of activities bound together by a diverse fabric of cooperation mechanisms which is flexible and dynamic to change. The existing legal framework for international cooperation by being broad absorbs the tectonic shifts in technology, geopolitics and geo-economies. Scholars argue that international cooperation should not be forced upon countries, because without shared interests' cooperation cannot be fruitful.[89] This, in itself endorses the wisdom of keeping space legislation broad-based and open. It is a convergence of interests that now promotes international cooperation as never before and the existing broad legal framework serves it well. Consequently, it would be reasonable to infer that it can sustain and weather the commercial onslaught without any drastic changes.

However, it needs to be borne in mind that this has led to incidents like India inadvertently launching satellites of foreign companies (Swarm) that were not authorized by their country of origin.[90] This brings to fore the perils of unregulated commercial expansion. The convergence of interests needs to be streamlined to ensure harmonious advancement of all parties. The general obligations to cooperate harmoniously would need to be supported by specific obligations. These specific obligations emerge when viewed in the backdrop of a wide interpretation of peaceful use of space. While the space treaties, in letter, are restricted to states, they are not so in spirit going by the numerous references to 'all mankind'. Consequently, the obligation of peaceful

[87]European Union, "International Aspects", online: https://ec.europa.eu/growth/sectors/space/international-aspects_en, (accessed on 02 February 2018).

[88]Wilfred Jenks (1965), at 234.

[89]Benkö and Schrogl, *International Space Law in the Making*, *supra* note 60 at 213.

[90]Henry (2018).

use is not restricted to states, but extends to all mankind. This expanded coverage imposes a moral obligation on all parties, including commercial and private players to ensure peaceful use of space. Adherence to the moral obligation does not resolve the issue but certainly provides a common standard for convergence of interests and harmonious advancement of all parties.

3.4 Conclusion

International cooperation in space has come a long way since space law was formed in the 1960s when commercial activities were opposed in both letter and spirit.[91] In negotiating the Outer Space Treaty, the United States supported involvement of private players; but this proposal was opposed by the Soviet Union which wanted only States to undertake space activities. Ultimately, Article VI of the Outer Space Treaty was drafted to allow private activity in outer space on the condition that the appropriate State exercises authorization and continuing supervision over is its nongovernmental entities.

The formula has apparently worked well in view of the vast panoply of cooperation mechanisms in space. There exists enormous diffusion of technology today and the geo-political, geo-economical landscape is far different from the 1960s when the law was drafted. Cooperation, whether in the format of a multilateral forum promoting discussion of contemporary issues and development of treaties, or in the format of an international organization operating satellites, has always been a response to technical, scientific, economic and political change. As the change gathers pace and issues like private appropriation of celestial resources like asteroids come up, the lofty ideals originally envisaged would be challenged and tested in many ways. It remains to be seen whether the existing legal regime would be resilient enough to absorb and adapt to that change.

However, at present, the existing legal system works and is responsive to change. The idea of benefit appeals to common interest of state, non-state actors and as the number of stake holders and interests in space rise, the possibility of converging on common interest can only possibly rise.

[91]The Soviet Union was initially opposed to commercial participation in space. Subsequently, it agreed that "it would be possible to consider the question of not excluding from the declaration the possibility of activity in outer space by private companies, on the condition that such activity would be subject to the control of the appropriate State, and the State would bear international responsibility for it." *Committee on the Peaceful Uses of Outer Space*, 22d mtg., UN Doc A/AC.105/PV.22 (13 September 1963) at 23. See also von der Dunk (2008), 531 at 532.

References

African Union (2017) African space strategy: towards social, political and economic integration. In: HRST/STC-EST/Exp./16 (II), p 23. https://www.au.int. Accessed 14 Jan 2018 (October)

Benkö M, Schrogl K-U (eds) (1993) International space law in the making: current issues in the UN Committee on the peaceful uses of outer space. Editions Frontieres, Paris, p 213

Benkö M, Schrogl K-U (2000) Space law at UNISPACE III: achievements and perspectives. Z Luft Weltraumrecht 49(1):74

Bloomfield LP (1965) Outer space and international cooperation. Int Organ 19(3):603–604

Bourley M (1988) Space commercialization and the law. Space Policy 4(2):131

Brunner C, Soucek A (2011) Outer space in society, politics and law. Springer, New York, pp 326–327

Chaldek J (2017) Outposts on the frontier: a fifty-year history of space stations. Nebraska Press, London, pp 2, 9–10

Conforti B, Labella A (2012) An introduction to international law. Brill Publishers, Leiden, pp 35, 42–43

Davenport C (2018a) The Trump administration wants to turn the International Space Station into a commercially run venture, NASA Document Shows. The Washington Post (11 Feb)

Davenport C (2018b) NASA's new administrator says he's talking to companies about taking over operations of the International Space Station. The Washington Post (5 June)

Del Rio Vera J (2017) Space sustainability in the 21st century. Paper presented at the clean space industrial days conference, 25 Oct 2017, ESTEC, Noordwijk, the Netherlands. https://indico.esa.int/indico/event/181/session/3/contribution/120/material/slides/0.pdf. Accessed 28 Oct 2017

Dolzer R (1985) International cooperation in space. Z ausländisches öffentliches Recht Völkerrecht. Heidelberg J Int Law 45:527. http://www.zaoerv.de/45_1985/45_1985_3_k_527_546.pdf. Accessed 26 Dec 2017

Galloway JF (2008) Revolution and evolution in the law of outer space. Nebr Law Rev 87(2):516–517

Georgion A (2018) NASA might be selling the ISS to private companies. Newsweek (6 June)

Ghali BB (1994) International cooperation in space for security enhancement. Space Policy 10(4):265

Guterres A (2018) Secretary General's video message for UNISPACE+50. UNISPACE. https://www.un.org/…/secretary-generals-video-message-unispace50-high-level-segme. Accessed 17 July 2017 (20 June)

Hahn HJ (1997) International organizations, resolutions. In: Encyclopedia of public international law, vol 2, p 1334

Henry C (2018) FCC issues warning in wake of Swarm's Unauthorized Launch. Space News. https://spacenews.com/fcc-issues-warning-in-wake-of-swarms-unauthorized-launch/. Accessed 16 Apr 2018 (13 April)

Jakhu R (2005) Legal issues relating to the global public interest in outer space, p 6. https://drum.lib.umd.edu/bitstream/handle/1903/7916/jakhu.pdf;sequence=1. Accessed 26 Dec 2017 (October)

Jakhu RS (2009) The effect of globalisation on space law. In Hobe S (ed) Globalisation—the state and international law. Franz Steiner Verlag GmbH, Stuttgart, pp 71–77

Jakhu R (2010) National regulation of space activities. Springer, Dordrecht

Kendall D (2018) Reflections on the safety, security and sustainability of outer space: the United Nations. In: Canadian Air and Space Institute ASTRO-18 conference, 15–17 May 2018

Lopez LD (2016) Regional space cooperation: spotlight on Latin America. Secure World Foundation. https://swfound.org/media/205496/2016-ldl_regionalspacecooplecture_f.pdf (March)

Lyall F, Larsen PB (2009) Space law: a treatise. Routledge, New York, p 44

Siddiqi AA (2015) Science, geography and nation: the global creation of Thumba. Hist Technol J 31(4):420

Suy E (1977) The meaning of consensus in multilateral diplomacy. In: Akkerman RJ, van Kreiken PJ et al (eds) Declarations on principles: a quest for universal peace. AW Sijthoff Publishing Company, Alphen aan de Rijn, p 260

US Department of State (1998) Treaties and other International Acts series 12927, Space Station Agreements between the USA and other Governments. https://www.state.gov/documents/organization/107683.pdf. Accessed 12 Nov 2017 (29 Jan)

von der Dunk F (2008) Report of the 3rd Eilene M. Galloway symposium on critical issues in space law—Article VI of the outer space treaty: issues and implementation. In: 51 Proceedings of the international institute of space law, pp 531–532

von der Dunk FG (2014) International space law. In: von der Dunk FG (ed) Handbook of space law. Edward Elgar, Cheltenham, p 58

Voronina A (2016) The how's and why's of international cooperation in outer space: international legal forms of cooperation of states in exploration and use of outer space. Thesis, University of Nebraska

Wilfred Jenks C (1965) Space law. Praeger, New York, p 234

Chapter 4
Legal Options to Address the Challenges and Opportunities in Small Satellites

Abstract The surge in mega-constellations of small satellites is an aspect which is both unprecedented and unexpected. It carries opportunity and challenge in equal measure. Consequently, it would be essential to regulate this area to ensure optimal and judicious exploitation. This chapter analyses the ability of international law to address common challenges arising out of the surge in small satellites and how convergence on these challenges carries with it the potential for sustainable development of space.

4.1 The Small Satellite Surge

The era of small satellites is back. They are cheap, expendable and practical in many more ways than one. Despite the first satellite Sputnik being a small satellite and sporadic advances by the Soviet Union in the 1960s; it is only due to recent advances in modern technology and miniaturization that they have become practical and popular. For well around four decades, from 1960 to 2000, there were very few small satellites, even during the years 2000–2012, the total number of small satellites launched were barely in the ranges of 20–25. The figures suddenly shot up to 92 small satellites in 2013, the next year it reached 158 and the numbers peaked at 300 in 2017.

4.1.1 Mega-Constellations of Small Satellites

The figures for the present are in hundreds, however, future trends indicate small satellites in thousands. For instance, in March 2018, the US Federal Communications Commission approved SpaceX's application to launch 4425 low-Earth orbit satellites for high speed broadband across the world.[1] Two satellites of this mega constellation have already been launched. These 4425 satellites are the first phase of nearly 12,000 satellites envisaged. The remaining 7518 satellites would follow later.[2]

---

[1] See Brodkin (2018).

[2] Grush (2018a).

K. K. Nair, *Small Satellites and Sustainable Development - Solutions in International Space Law*, SpringerBriefs in Law, https://doi.org/10.1007/978-3-030-18620-3_4

Table 4.1 Trends in small satellites[a]

Year	2000–12	2013	2014	2015	2016	2017
Satellites	20–25	92	158	131	101	300

[a]Table created by collating data from *Jonathan's Space Report* online: http://www.planet4589.org/space/jsr/jsr.html; *Space Security Index 2017*, online: http://spacesecurityindex.org/; and Space Works, "Nano/Microsatellite Market Forecast 2017", online: https://www.spaceworkscommercial.com/wp-content/uploads/2018/01/SpaceWorks_Nano_Microsatellite_Market_Forecast_2017.pdf

Table 4.2 Small satellite mega-constellation trends

Company	No. of sats	Orbit	Mass (kg)	Frequency	Remarks
Space-X	4425	1100–1325 km 83 Planes 53–81° Incl	100–500	Ku and Ka	First two satellites in orbit. (22 February 18)
Steam/Norway	4257	LEO, 43 Planes	–	Ku and Ka	–
MCSAT	4000	LEO	–	–	
1 Web	900	1200 km 18 Planes 87.9° Incl	175–200	Ku	First to register fre-quency with FCC
Boeing (Viasat)	2956	1200 km 45–88° Incl	–	V band	–
Comstellation/Canada	794	LEO 12 Planes	–	Ka	–

Similarly, Norway has filed with the International Telecommunication Union (ITU) for registration of its Steam network of 4257 satellites, France has filed for its 4000 satellite MCSAT constellation and Canada has filed for a 794-satellite constellation.[3] Quite clearly, the trendlines indicate a rise in small satellites from figures of tens to hundreds and now thousands.

Table 4.1 serves to illustrate the trend with regards to small satellites until the year 2017.

The trends in mega-constellations of small satellites are unprecedented as shown by Table 4.2.[4]

[3]de Selding (2015).

[4]Table created by collating data from various sources including Larbi et al. (2017). See also *Small Satellite Market and Report*, Orange Silicon Valley (August 2016).

Table 4.3 Small satellite classification

	Classification of small satellites by mass	
Satellite class	Mass range	Functionality
Femtosatellite	10–100 g	In swarms
Picosatellite	100–1000 g/1 kg	In swarms
Nanosatellite	1–10 kg	Individually & in groups
Microsatellite	10–100 kg	Individually & in groups
Small satellite	100–500 kg	Individually & in groups

4.1.2 Small Satellite Classification and Functions

Across the globe, more and more agencies (national, multinational and private) are putting small satellites into orbit since they are cheap, practical and expendable. Their size is small as compared to conventional satellites. For example, conventional communication satellites in GEO range from sizes as big as a School bus to a lawn mower with solar panels extending well beyond their main bodies. However, the word 'small satellites' is suggestive only of its size and not role since modern technology allows smaller payloads to have the same and, in some cases, better efficiency than large satellites. The sizes in small satellites vary and in order to be on a common grid, it would be essential to briefly examine the terms as they are known in common parlance and the same is shown in Table 4.3.[5]

In the above classification are Cube-sats, a subcategory of Nanosatellites that weigh up to 1.33 kg and are 10 cm in all three dimensions. They are cheap at barely $50,000[6] and very popular with universities and companies.[7]

With regards to uses, the initial utility of small satellites was largely confined to environmental observation, scientific tests and communication relay, the utility today is more varied, multifaceted and complex. They can be put to a variety of civilian uses like Earth observation (IMS-1), disaster monitoring (DMS), education (Annasat, Ardusat), astronomy (Brite-PL) as also military applications like the SENSE-1 of the US Air Force etc. These are low-mass and low-cost platforms that can be sent into orbit for much less than a few million Dollars and are an attractive option for space faring as also non-space faring nations, corporations, educational institutions as also individuals. Unlike a medium or large satellite that is difficult to make and even more difficult to launch, small satellites provide an easier and affordable alternative. Since they are small, they are easier to launch and one is now witness to hitherto unheard missions like a single launch by Orbital Sciences that put a record 29 satellites into Low Earth Orbit in November 2013.[8] Thirty hours later, *Kosmotras*, a Russian joint-

[5]There exists no commonly accepted standard of small sat size. However, most literature takes the NASA standards into account. See Mabrouk (2015).

[6]David (2017).

[7]Erik Kulu, "Nanosatellites by Organisation", *Nanosatellite Database* online: http://www.nanosats.eu/ (accessed 19 April 2018).

[8]Wall (2013).

venture, carried 32 satellites into a similar orbit.[9] Then, in January 2014, Orbital Sciences carried 33 satellites up to the International Space Station (ISS), where they were cast off a month later.[10] A similar logistic resupply mission to the ISS was by Space-X that envisaged placing in orbit 104 'Sprites', not much larger than postage stamps that contain all the basic elements of a satellite like radio, solar cells, aerials and other instruments.[11] However, due to a fault the mother ship failed to deploy and they burnt on re-entry.[12] The above notwithstanding, across the world, a variety of small satellite projects are on the anvil. Put briefly, as the mass is small, launch is easier and as the uses and users of space rise, one can only expect the proliferation of microsatellites to rise exponentially and herein lie the dangers of unregulated growth and expansion.

The dangers of such growth are evidenced at both domestic and international levels. At domestic levels, states have their own individual standards of licensing leading to a variety of issues. For example, US standards are quite stringent in that the launch of small satellites are regulated under the 1984 Commercial Space launch Services Act. The launch licenses are issued by the Office of the Secretary of Transportation, which has delegated its authority to the Office of Commercial Space Transportation (AST) within the Federal Aviation Administration (FAA). A license is issued after a thorough safety and mission review is conducted by various US agencies. Further, before issuing a license, AST must ensure appropriate licenses have been obtained from the Federal Communications Commission (FCC), the Department of Commerce, Department of Defense and other relevant agencies. The license issued is again subject to a variety of conditions.[13] By contrast, India's launch standards are less stringent. A customer seeking launch services only needs to enter into a contract with the Antrix Corporation, which is the commercial arm of the Indian Space Research Organization. The authorization is provided through the Department of Space taking into account international treaty obligations and domestic considerations. India's launch services are simple and expeditious. Most small satellites look for launch at the earliest opportunity. This, to an extent, perhaps, explains why a US start up called Swarm technologies launched its small satellites on an Indian PSLV despite being denied authorization by the US FCC. Later reports indicate the act was not malafide and was the result of a series of mistakes arising largely out of the complexities in procedures.[14] Quite clearly, there exists a need for standardization of regulations all over the world to avoid the chaos, keep all players on board and simplify procedures specifically in the context of small satellites.

---

[9]Clark (2013).

[10]NASA (2014).

[11]AMSAT-UK (2014).

[12]Whitman (2017).

[13]Jakhu and Pelton (2014).

[14]For details, see Grush (2018b).

4.2 The Impact of Lack of Specific Small Sat Regulation

The trends indicate a rise in swarms of small satellites which is astonishingly rapid, unprecedented and unruly as of now. It is a revolution by itself and far outstrips the pace of review, reform and regulation possible by international bodies like the United Nations (UN) or the International Telecommunication Union (ITU) and herein lies the danger of continuing with *lex lata*, or the law as it exists.

The issues are manifold; there exists no legal definition of a small satellite, registration issues related to swarms of multi-agency satellites are vague. Also, spectrum, frequency, slot allocation issues are already heavily contested and in case of swarms of small satellites represent a potential nightmare. Further, these rising numbers translate into overcrowding in useful orbits and increased potential for conflict over scarce orbital resources. It also adds to space debris. Apart from the obvious, the inherent versatility of small satellites endows them a variety of Anti-Satellite roles. There is no verification mechanism as of date and the potential for interference; intentional or unintentional, harmful or otherwise is immense. With no semblance of global space traffic management in place, the situation is precarious. In fact, the possibility of mutual suspicions and rivalry snow-balling into actual conflict in space is no longer remote. The magnitude of the problem is enormous, global in scope and unless steps are taken today to regulate the plethora of problems likely with the profusion of small satellites, all of humanity tomorrow might lose access to space.

4.2.1 *Rise in Frequency Allocation and Allotment Complexities*

On the brighter side, the democratization of space capabilities is good and desirable; however, the manner of the present growth and proliferation of small satellites is unruly at best and fraught with dangers for all of humanity. The rise in swarms of small satellites is astonishingly rapid, unprecedented and increases the complexities and challenges in general terms for regulatory bodies like the UN, and in specific aspects for the ITU. The challenges are increasingly manifest with every constellation launch and the existing resources of the ITU are under equipped to deal with the issue.

To elaborate, the ITU has played a significant part in regulating radio frequencies since 1959. However, it needs to be borne in mind that satellite transmissions are only one of the many kinds of international communications regulated by the ITU. As a matter of fact, ITU treats space radio as just another aspect of the general regulations on use of the radio spectrum that finds its place in the overall definition of telecommunication as:

> Any transmission, emission or reception of signs, signals, writings, images and sounds or intelligence of any nature by wire, radio, optical or other electromagnetic systems.[15]

[15]ITU Constitution, *supra* note 71.

The issue of small satellite now demand specific treatment in view of the large numbers involved. Satellites in thousands would require assigning of particular frequencies and orbital slots. As in case of a normal satellite, these small satellites would also have to follow the procedural chain and the ITU can be expected to be swamped by the number of notifications of assignments imminent. Potential interference with assignments already registered on the MIFR would need to be checked even where a notifying state has already achieved coordination with other states. This demand enormous coordination by the ITU to ensure requisite procedures are fulfilled and no harmful interference of frequencies take place. Radio frequencies and orbital positions in GEO are limited natural resources shared among various radio services and all countries. Consequently, the ITU has a complex, elaborate procedure governing its allotment.[16] Put briefly, the ITU satellite coordination process to avoid harmful interference may take anywhere between two to seven years before actual operation and hence the complexities in case of mega-constellations of thousands of satellites can well be imagined. On the other hand, the present mega-constellations are all being launched by developed nations and consequently this would bring in the question of equity with respect to emerging and developing nations. In order to ensure equitable access to radio frequencies and orbital positions, the ITU Constitution specifies that the special needs of developing countries are to be taken into account and this enables a priori allotment of radio frequencies and orbital positions. However, this clause is limited only to the Geo stationery Earth Orbit (GEO) and not the LEO where the mega constellations are planned. In order to ensure the spirit of the principle of equity in the ITU constitution as also the space treaties is not violated, it would be essential to declare LEO as a limited resource and bring it within the scope of equitable distribution of space resources.

4.2.2 *Space Debris and Related Issues*

Radio frequency regulations in outer space are nothing more than one of the instances that constitute a grey area. A variety of other issues abound; there exists no definition of a small satellite, the registration issue with respect to swarms of multi-agency satellites are vague, spectrum, frequency and orbital slot allocation issues in LEO are a potential nightmare. Apart from the above, increasing number of satellites translates into overcrowding in useful orbits and ominously increased debris. The amount of junk orbiting in outer space certainly inspires very little confidence in any manner of environmental damage mitigation or sustainable development in outer space. The orbital debris hazards of the potential swarms of satellites once they have outlived their utility are manifold and yet to be comprehensively understood for environmental mitigation to be put in place.

As per international space law, particularly the Outer Space Treaty-1967, space is global commons, free for exploration and use by all, as long as the use as per Article

[16]For details on allocation and allotment procedures, see Jakhu (2016).

VI of the OST is responsible, authorized, supervised by the state and does not cause harmful interference. In case of conventional satellites that involve huge costs in terms of money, effort, time and other resources the levels of responsibility tend to be inherently high since the stakes are high. However, in case of small satellites the converse applies. They are cheap to produce, launch and hence expendable. Long term returns on investment, functionality or responsible use are not priority areas in case of small satellites. As Steven Freeland points out:

> the existing legal framework was not designed with small satellite technology specifically in mind... activities involving small satellites typically fall within the scope of Article VI of the OST. This in itself is not surprising – what is, however, is that this is not necessarily understood by the users of small satellites, particularly with respect to experimental projects...Many experimental satellite programs have been exactly that – experimental. They have often utilised existing off-the-shelf components, and the expectations of mission success for any significant period of time have not necessarily been high. It is fair to say that such circumstances give rise to lower perceptions of risk and a high tolerance towards failure. For many such programs, at least in the relatively early phases of small satellite development, the process has largely been about the journey to space rather than the delivery of services.[17]

Lending credence to the above statement is data related to nanosatellites; most of which are launched by the private industry and academia for experiments rather than long term use.[18] As clearly borne out by Fig. 4.1, the missions in most cases deal with proving concepts, technology demonstrations, demonstration of low-cost access to space and scientific research among other things.[19]

Space activities, in case of most new space actors, particularly those dealing with small satellites are confined to short term experiments rather than conventional space applications like communication, navigation that typically last for years.

The aspect about the journey being more important than the service is evidenced in a variety of cases including university experiments, first time launches symbolizing national pride among other short term uses. For instance, Colombia's Sergio Arboleda University launched Libetad-1 for transmission of one stanza of their national anthem.[20] These are legitimate endeavors that reflect the aspirations of states and private parties and hence in these cases the journey is far more important than the service since it symbolizes the opening up of new vistas to new players. Secondly, experimentation is the crucible for scientific advance and technology maturation. An aspect guaranteed under various clauses of the OST including the preamble and especially Article 1 OST which states:

> The exploration and use of outer space, ...shall be carried out for the benefit and in the interests of all countries, irrespective of their degree of economic or scientific development

[17] Freeland (2015).

[18] Erik Kulu, "Nanosats and Cubesats Database" online: https://www.nanosats.eu (accessed 2 June 2018).

[19] UN Office of Outer Space Affairs (UNOOSA), "On Line Index of Objects Launched into Outer Space", online: http://www.unoosa.org/oosa/osoindex/search-ng.jspx?lf_id= (accessed 6 June 2018).

[20] "Colombia Launches First Satellite" *Space Daily.Com* (April 2007) online: https://www.spacedaily.com/reports/Colombia_Launches_First_Satellite_999.html (accessed 10 June 2018).

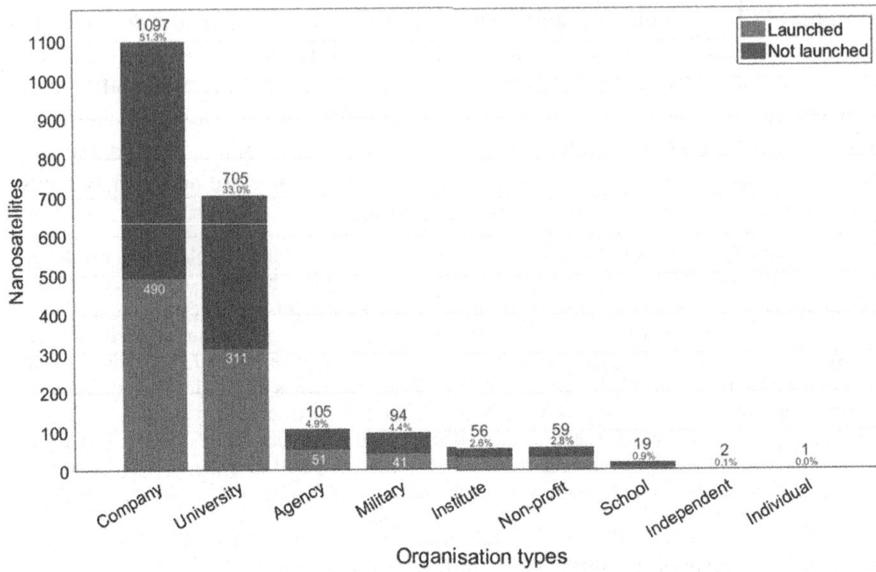

Fig. 4.1 Number of nanosatellites by organisation (Kulu, *supra*, note 20)

> …There shall be freedom of scientific investigation in outer space, including the Moon and other celestial bodies, and States shall facilitate and encourage international cooperation in such investigation.[21]

Thus, short term experimentation in space is not only a legitimate activity, but an activity that has to be supported and encouraged. However, this leads to the question of what happens after the short-term use. Unlike large satellites that have life spans of 10–15 years and carry de-orbit fuel reserves, small satellites have short life spans of barely 2 years and have little fuel left after station keeping of two years to de-orbit and remain in space as debris. A significant number of CubeSats are dead on arrival and some do not last beyond a few days or weeks.[22]

No hard laws exist on debris mitigation and attempting to introduce such laws has led to conflicts of rights amongst the stakeholders. Perhaps, the answer lies in states facilitating and encouraging international cooperation, an aspect clearly borne out in Article 1 OST but not in practice. There exists duplication and repetition in many experimental endeavours; for instance, many experiments deal with proving amateur radio services, gaining basic knowledge and skills, basic imagery which could be avoided if the experimental results are shared under the aegis of international cooperation in space.

Similarly, Universities across the world could collaborate on projects, so that experimental results of a general nature are disseminated and the same experiments

---

[21] *Treaty on Principles Governing the Activities of States in the Exploration and Use of Outer Space, including the Moon and Other Celestial Bodies*, 27 January 1967, 610 UNTS 205 [OST], Article 1.

[22] For details, see Venturini (2017).

are not repeated. A common forum of Universities across the world for proposing, vetting, financing, facilitating and disseminating knowledge related to space experiments institutionalized under an international body like the United Nations may be far more viable, purposeful than disorganised, individual launches from all over the world. However, these regulations may not comprehensively address the issue of established players launching constellations of mega satellites in space fór commercial space applications. The need to balance aspiration with regulation is paramount for harmonious growth of new players and sustainable development of space.

4.2.3 National Security Issues

Small satellites present opportunities and challenges in equal measure. On one hand, student endeavours like Annamalai University's 50 kg microsatellite called Anusat as also IIT Kanpur's four kg nano-satellite Jugnu, University of Alberta's, AlbertaSat and numerous other university endeavors do serve to ignite young minds and encourage innovation.[23] They also enable cost-cutting and large profit margins. However, if one goes beyond the academic, economic factors and takes into account the factors on national insecurity presented by small satellite proliferation, an ominous picture unfolds.

On one hand, the possibility of space capabilities being misused to support criminal, terrorist activity is no longer remote. The relative obscurity of SATCOM is known to facilitate money laundering and narco-trafficking among other activities. On the other hand, a variety of Anti-Satellite roles can be envisaged for small satellites with the right kind of equipment in place. There exists no verification mechanism as of date and the potential for interference; intentional or unintentional, harmful or otherwise is immense.

Equally immense is the possibility of suspicious maneuvers by satellites (perceived suspicious or otherwise) snowballing into bitter acrimony. Maneuvering small satellites could be used for on-orbit inspection, repair etc. and they could also be used to collide into other satellites or even carry lasers or other directed energy weaponry to burn satellite optics, solar panels and other space craft material.

The possibilities and scenarios are manifold and the sheer number, flexibility of roles and lack of homogeneity in small satellites as also the challenges of tracking small satellites demand regulation of activities of small satellites. The magnitude of the problem is enormous and unless steps are taken today to regulate the plethora of problems likely with the profusion of small satellites in usable earth orbits, all of humanity tomorrow might lose access to space.

[23]Cummings (2014).

4.3 The Regulations Enveloping Small Satellites

The legislation applicable to outer space comprises of the principles and rules of General International Law, the Outer Space Treaty-1967 (OST), Regulations of the ITU, international agreements and soft-law like the IADC guidelines, UN Resolutions. In general, all the rights and obligations applicable to large satellites apply to small satellites. Existing legislation does not distinguish between small and large satellites, perhaps because a small satellite surge was never anticipated when the OST was created in 1967. Thus, all small satellites, regardless of their mass, size, function are considered space objects governed by existing international legal guidelines.

4.3.1 Definitional Issues with Regards to Satellites

To begin with, the *lex specialis* of outer space comprising of the five space treaties do not make any mention of the word "*satellite*". The operative word across the preamble and text is initially confined to 'objects' in the treaties and later progresses to 'space objects' by the time the Liability Convention was inked in 1972. As per Article 1(d) of the Liability Convention:

> The term 'space object' includes component parts of a space object as well as its launch vehicle and parts thereof.[24]

Even in this case, the definition is preceded by a qualifier that the term is for the purpose of the convention. Evidently, the definition is general at best in that it covers launch vehicles which are not satellites by any yardstick and mentions space objects which may include satellites, space probes, space stations, satellite antennae, solar panels etc. The term first finds mention a decade later in the principles governing the use by states of artificial earth satellites for international direct television broadcasting (1982). However, even here, the term is not defined as such.

On the other hand, the ITU regulations, in particular the Radio Regulations[25] sets out the terminology and are helpful with regards to definitions. The RR 1.64 defines a 'space station' as being 'located on an object which is beyond, is intended to go beyond, or has been beyond the major portion of the Earth's atmosphere' and RR 1.178 defines a 'space craft' as a man-made vehicle which is intended to go beyond the major portion of the earth's atmosphere. These definitions also suffice only to provide a broad guidance and hence it may be reasonably inferred that the term has been left broad to encompass the vast panoply of manmade objects possible in space and to enable flexibility in application of regulation. In view of the foregoing, it may be safe to infer that the normative interpretation of a satellite as a space object

---

[24] *Convention on International Liability for Damage Caused by Space Objects*, 29 March 1972, 961 UNTS 187 (in force 1 September 1972).

[25] *Radio Regulations of the Radio Regulations Board*, International Telecommunication Union, compiled in *Radio Regulations*, (Geneva: International Communications Union, 2012) at 1.

enabling space applications like observation, communication, navigation and other support functions is the most suitable. However, with respect to small satellites, the inherent flexibility in roles possible distinguishes its character from a normal big satellite. For instance, a big communications satellite could be purposefully used only in the GEO for satellite communications since its movement is synchronized with that of the earth, it has an ITU assigned orbital slot, frequency and is a costly asset. However, a small satellite in LEO has both freedom of movement as also mission in that it can operate anywhere in LEO, it can act both as radio relay satellite and also carry a payload (optical camera, IR sensor) for observation/imaging and could also be used to collide deliberately with other space objects. For the above reasons, in the eventuality that small satellites in thousands proliferate, it would be essential to treat them separate from traditional big satellites.

4.3.2 Small Satellite Issues in the Outer Space Treaty

A satellite qualifies as a space object and hence the space treaties apply across the board. However, in specific terms, certain provisions apply in greater measure and the same are discussed briefly. Firstly, under the provisions of Article VI,[26] OST, states bear international responsibility for national activities in space whether such activities are carried on by governmental or non-governmental entities and the activity shall require authorization and continuing supervision by the state. Thus, all small satellite activity effectively is covered under the ambit of Article VI, and yet there have been instances of unwillingness on part of states to do so. The Dutch Government was reportedly unwilling to require the licensing of cube-sats on argument that cube-sats were not 'active' because they lacked any propulsive or controlling element and hence not an 'activity'.[27]

Equally significant is the aspect that when a space object is launched into Earth orbit or beyond, a state is required to register it with the Secretary-General of the UN under the Registration Convention-1975 (RC) or UNGA Res 17218(XVI).[28] Also, as per Article VIII of the OST, a state party on whose registry an object launched into outer space is carried shall retain jurisdiction and control over such object. These regulations are designed for individual space object; it is with "*a space object*" in mind and not a deluge of small satellites or mega-constellations of thousands of satellites that the regulations were drafted. The minimum data required for registration has been laid down in Article IV (1) of the RC-1975 and includes elaboration on a variety of aspects ranging from functions to basic orbital parameters that in terms of conventional big satellites are largely static but no so in case of small satellites. The configuration, orbital integrity of the constellation can also be

[26]OST, *supra* note 24, Article VI.

[27]For details, see Masson-Zwaan (2012, 2016), Palkovitz and Zwan (2015).

[28]*Convention on Registration of Objects Launched into Outer Space*, 14 January 1975, 1023 UNTS 15 (in force 15 September 1956) [Reg Conv].

expected to be fluid unlike a static single satellite and this would demand enormous coordination between the state registering body and the international registration body of UNOOSA. The existing procedures for registration would need to take into account not only the large numbers but also the multiple functions, the orbital parameters and other aspects. Integrating existing registered data on satellites with thousands of new small satellites, some of whom may not have a life beyond a few months may lead to needless confusion, duplication, mis-identification and may serve to defeat the basic purpose of identification of satellites.[29]

There also is the question of obfuscation of state responsibility and state of registry. Constellations when licensed and handled by a single state do not pose much of a problem, however, issues arise when the launch is undertaken by a different state, the operations are conducted by another and ownership rests with a third party. For example, in a recent instance, India launched four small satellites of US based Swarm Technologies despite the US Federal Communications Commission (FCC) rejecting the company's application for a license.[30] The indisputable fact is that the small satellites are in orbit. The legal question on dispute is on whose registry and to whom do these small satellites belong? Who is responsible for their continuous authorization and supervision and finally who assumes liability in case of damage? Evidently, there are no clear answers and the obfuscation demands clarification by legal review and reform urgently.

Additionally, a state is "*internationally liable*" for damage caused by a space object that it launches or procures the launching of or from whose territory or facility an object is launched. Article VII of the OST simply holds that a state that launches or procures the launch of an object in space is liable for any damage it causes. The principle is further elaborated upon in the 1972 Liability Convention (LC).[31] As per the LC, damage caused on the surface of Earth or to aircraft in flight involves absolute liability,[32] while damage to objects in space is fault-based liability. Mega-constellations of satellites indicate mega-numbers and to the question of as to whether an increased number of space objects indicate increased liability concerns in space, the answer is unequivocally in the affirmative. Unlike normal liability cases involving motor vehicles, aircraft on earth, cases involving space objects liability in space till date have been few and far in between primarily because the satellites were large, identifiable and less in numbers. The above would most likely change in view of the small satellite surge. The numbers would rise as would the risks, the costs would be driven proportionately higher and indirectly impact the aspirations of developing nations and private parties low on finances. An unintended consequence of this would be a conflict with the hallowed principle of equity and the belief that the 'exploration and use of space shall be carried on for the benefit of all peoples irrespective of the degree of their economic or scientific development' which is enshrined across

[29] *Ibid.* Preamble.

[30] See Sheetz (2018).

[31] *Convention on International Liability for Damage Caused by Space Objects*, 29 March 1972, 961 UNTS 187 [LC].

[32] *Ibid.* Article II.

the space treaties. An entire spectrum of liability measures is conceivable in that some states may waive off liability to promote research and development, some may simply not make liability insurance a mandatory requirement and some may attach high standards of liability or may attach different standards of liability for big and small satellites.

With regards to liability, states may be expected to do what is in their best interests as also that of its citizens. However, as borne out by Jakhu, the principle of liability in space holds much more than simply liability. It includes responsibility also:

> It is only in the English version of the treaty that a distinction is made between responsibility and liability. The texts of the Treaty in the Chinese, French, Russian and Spanish languages which are equally authentic make no distinction between responsibility and liability. Therefore, a liability claim for compensation can also be made under Article VI of the OST.[33]

Thus, it may be surmised that the option of states waiving off liability simply does not exist. A state party to the treaty would have to mandatorily accept responsibility and liability for small satellites of the state or its private parties and this is a burden some developing states just would not be able to bear or might not find cost-effective. Complicating issues is the aspect that apportioning of blame for fault-based liability in space requires observation, tracking of space objects by a Space Situational Network (SSN) capability that at present rests only with the US. The SSN is by no means comprehensive and there have been collisions in space which the SSN neither observed nor averted. The SSN, at best is for ensuring safety of US space objects and not the world at large. Consequently, in practical terms, small satellites present a small, at times undetectable cross-section and so the likelihood of clear, convincing fault-based liability is remote. In case of nations as claimant states with some SSN capabilities of their own, it would be difficult, if not impossible to establish fault as to who caused the damage to their satellite. This would be especially so in case of developing nations operating small satellites with no SSN capabilities at all. The complications of fault-based liability in space for individual big satellites are high, they can be expected to compound in case of thousands of small satellites. A balance would hence need to be found between legitimate aspirations for scientific research, development and sustainable use of space.

4.4 Small Satellites, Space Debris and Sustainable Development

The sheer volume of satellites involved in mega-constellations brings into play the aspect of space debris. Unless measures of Active Debris Removal (ADR) and Remediation are undertaken the already congested useful belts in LEO can be expected to become unusable by all humanity in the future. International space law does not

[33] Jakhu and Pelton, *supra* note 14 at 63.

specifically address the issue of space debris creation though it does deal with the consequences if the damage is caused by space debris. However, Article III of the OST specifically draws all space activity within the ambit of General International Law stating:

> State parties to the Treaty shall carry on activities in the exploration and use of outer space, including the moon and other celestial bodies, in accordance with international law, including the Charter of the United Nations, in the interest of maintaining international peace and security and promoting international co-operation and understanding.[34]

Consequently, the principles of international law, particularly those relating to the environment apply and are explored below.

4.4.1 Small Satellites and International Environmental Law

State obligations with regards to the environment have been recognised since the *Trail Smelter Arbitration*[35] which articulates the duty of a state to not permit the use of its territory to the detriment of another state. The 1992 UN Framework Convention on Climate Change (UNFCCC) reflects the same principle with its Preamble recognizing that states are responsible "*to ensure that activities within their jurisdiction or control do not cause damage to the environment of other states or of areas beyond the limits of national jurisdiction*".[36] The aspect that states have a responsibility to ensure activities within their control do not cause damage to the environment beyond their national jurisdiction further resonates in Principle 21 of the Stockholm Declaration of 1972.[37] Similar language is found in a number of later agreements and declarations. These declarations are largely aspirational and non-binding; however, they do indicate both a recognition of the issue and a willingness to cooperate in resolving the issue. Apart from the declarations, an inclination towards the general international duty for preservation and conservation of the environment is evident in the Advisory Opinion of the International Court of Justice (1996) on the Legality of the Use by a State of Nuclear weapons in Armed Conflict wherein the ICJ stated:

> The Court also recognises that the environment is not an abstraction but represents the living space, the quality of life and the very health of human beings, including the generations unborn. The existence of the general obligation of states to ensure that activities within their jurisdiction and control respect the environment of other states or of areas beyond national control is now part of the corpus of International Law relating to the environment.[38]

---

[34] OST, *supra* note 24, Article III.

[35] *The Trail Smelter Arbitration Case* (*US vs. Canada*) 1938/41 3 RIAA 1905; (1939) 33 AJIL 182; (1941) 35 AJIL 684.

[36] *UN Framework Convention on Climate Change*, UN Doc FCCC/INFORMAL/84;1771 UNTS 165; 1995 UKTS 28, Cm 2833; 1992 31 ILM 851-73 (1992).

[37] *Declaration of the UN Conference on Human Environment*, 11 ILM 1416 (16 July 1972), online: http://www.un-documents.net/unchedec.htm (accessed 24 July 2018).

[38] *Legality of the Use by a State of Nuclear Weapons in Armed Conflict*, Advisory Opinion, [1996] ICJ Rep 226 at para 29 (240).

Consequently, though outer space is beyond the limits of national jurisdiction, the obligations of international environmental law apply. And these obligations do not just apply post hoc. Francis Lyall and Paul B Larsen point out that the 'Precautionary Principle' argues in favour of giving a hard content to international environmental duties to avoid or prevent problems.[39] They contend that, it is better to take precautions which may not be needed, than to fail to take them and risk unfortunate consequences. In the context of mega-constellations, the above principle enables pre-emption and creation of regulation well in advance of the constellations being in place. Secondly, technical barriers and scientific uncertainty[40] are often cited as issues with regards to debris mitigation and remediation measures in formulating regulation. However, Principle 15 of the Rio Declaration (1992) makes it clear that:

> In order to protect the environment, the precautionary approach shall be widely applied by States according to their capabilities. Where there are threats of serious or irreversible damage, lack of full scientific certainty shall not be used as a reason for postponing cost-effective measures to prevent environmental degradation.[41]

Thus, there exists but very little reason to not apply the precautionary principle to space. Additionally, the aspect of states applying the precautionary approach as per their capabilities factors the considerations of developing nations. The threat to the environment of space due to mega- constellations is expected to be both serious and also irreversible to an extent. The overall effect of the above principle is that states can no longer rely on scientific uncertainty, technical challenges to justify a lack of action when there is enough evidence to establish the possibility of a risk of serious harm, even if there is no proof of harm.

It also needs to be borne in mind that law in general and in this case, international space law has evolved as per the context, needs and values. For instance, the early texts of the OST-67 and LC-72 base themselves on the Anthropocentric and Geocentric values. The sole reason for the protection of the extra-terrestrial environment in those times was the preservation of scientific opportunities. Later treaties like the Moon Agreement-79 (MA) evolved to emphasize preservation of the extra-terrestrial environment for inter-generational equity.[42] The concept of inter-generational equity introduced for the first time in the MA is aligned to the Rio declaration and geared towards inhibiting irreversible damage to the environment. In view of the mega-satellite surge and the need to preserve the environment of space for 'generations unborn' as the ICJ puts it, the concept of intergenerational equity assumes great significance. Its contemporary relevance could form the bedrock of binding legislation on space debris mitigation.

---

[39] Lyall and Larsen (2017), at 249.

[40] Tallis (2015), Walker et al. (2001).

[41] *The Rio Declaration on Environment and Development*, A/CONF.151/26, (2012), online: http://www.un.org/documents/ga/conf151/aconf15126-1annex1.htm (accessed 27 July 2018).

[42] *Agreement Governing the Activities of States on the Moon and Other Celestial Bodies*, UN Doc. A/34/664, New York, 5 December 1979 [Moon Agreement], Article IV.

4.4.2 Small Satellites and Debris Mitigation Guidelines

Space debris has been on the UNCOPUOS agenda since 1994 and the Inter-Agency Debris Coordination Committee (IADC) participates in and contributes to the UN space debris activities via the Scientific and Technical Subcommittee (STSC) of the UNCOPUOS. Comprising of thirteen space faring nations as members, it is established to exchange information on space debris research activities between its member space agencies. Additionally, the IADC recommends new opportunities for cooperation, serves as the primary means for exchanging information and plans concerning orbital debris research activities, and identifies and evaluates options for debris mitigation. Its findings took the form of the IADC Space Debris Mitigation Guidelines first in 2002, and subsequently updated in 2007.[43] The IADC uses surveillance data to analyze global trends in the adherence to mitigation guidelines and in February 2018 while it noted a satisfactory implementation of guidelines in GEO, in case of LEO, it observed that the current implementation levels were insufficient and no apparent trend towards better implementation was observed.[44]

With regards to mega-constellations, IADC issued its first statement in 2015 and followed it up in 2017 with additional recommendations on design and operation of such constellations. It seeks to reinforce the relevance of its existing space debris mitigation measures to constellation architectures and pursue a two-track approach to address this issue. First, on the basis of the outcome of initial reflections, it offers a number of preliminary qualitative observations that operators could consider in their conceptual design, and subsequently, as more substantive and comprehensive modelling data becomes available from the coordinated international studies, offer more detailed, quantitative guidance.[45] The IADC considers post mission disposal of satellites as one of the key drivers for the environmental sustainability of these missions.[46] Additionally, it lays down a variety of technical parameters for guidance ranging from design of the spacecraft, constellation to actual operation.[47]

However, it needs to be borne in mind that though the IADC provides technical recommendations on space debris, it is not a regulatory body. Therefore, the guidelines are not binding and without any enforcement mechanism. They are expected to be observed voluntarily. States and international organizations are expected to voluntarily take measures to ensure that the guidelines are implemented. In case of states, those incorporating these guidelines in their national mechanisms indicate that they have taken the best interests of all countries into account and are taking steps towards ensuring that space remains secure for all countries to use. This, by itself is significant for international cooperation as also debris management.

---

[43] IADC Steering Group (2017a), at 4.

[44] For details, see Ohnishi (2018).

[45] IADC Steering Group (2017b).

[46] *Ibid.*

[47] *Ibid.*

4.5 Recommendations *De Lega Ferenda*

In view of the foregoing, it is amply clear that mega-constellations of small satellites present a clear, present and imminent danger to sustainable use of outer space with no regulations in place to specifically address the issue. Even in terms of soft law, the IADC guidelines have been there for over a decade and as recently as year 2017, the IADC noted that there was no satisfactory implementation of its guidelines in LEO and future trends are also not promising. The call for a specific International Organisation to deal with the issue has been relentlessly pursued by Jakhu, Pelton and Dempsey[48] for almost a decade and has yielded but little. No international organisation on such grounds has been conceived and none is on the anvil. By contrast, the first small satellites of the mega-constellations have already reached orbit. Thus, it is essential that solutions be explored in light of past efforts, recent advances and the need to address issues with short term, intermediate and long terms solutions.

Since the Moon Agreement of 1979, no hard laws on space have come about. Secondly, specific to the issue of small satellites and amongst these, mega-constellations, the pace of technological advances far outstrips legal review, reform and hence it is imperative to prioritize the issue and explore solutions that appeal to the common vulnerabilities of most, if not all players. International law is consent driven and consent is far easier to obtain on common vulnerabilities than individual threats. Secondly, soft law initiatives like the IADC guidelines are more acceptable to states for various reasons including their non-binding nature. Attempting to address lacuna in soft law first is the easier, more acceptable and practical recourse and hence needs to be first dealt with. Keeping the above in mind, legal review and reform is proposed in the following areas.

4.5.1 Addressing Definitional Issues

The start point to legal review would be clarification of the definitional issues to enable clarity, legal certainty in the remaining discourse. As mentioned earlier, the term 'space object' has been defined in Article 1(d) of the LC and provides general guidance on the term. Secondly, the ITU RR 1.64 define a 'space station' and RR 1.178 defines a 'space craft'. It defines a 'space craft' as a man-made vehicle which is intended to go beyond the major portion of the earth's atmosphere. However, the IADC guidelines vide definition 3.2.1 also defines a 'space craft' as

> an orbiting object designed to perform a specific function or mission (e.g. communications, navigation or earth observation).[49]

This duplicity of definitions leads to needless obfuscation and uncertainty. The ITU definition is broad enough to serve the purpose of the ITU. However, the IADC

---

[48] Jakhu et al. (2011), at 27–31. Also, Jakhu and Pelton, *supra* note 183; and Jakhu et al. (2017).

[49] IADC *Space Debris Mitigation Guidelines*, IADC-02-01 (September 2007) at 5.

definition actually pertains to satellites in that it involves an orbiting object performing a specific function like communication, navigation, observation etc. What is alluded to is a satellite and the same may be clearly spelt out to avoid confusion and also because launchers have been separately defined in the IADC definitions. The IADC guidelines by virtue of being soft law are more amenable to modification than the ITU RR. Secondly, the context of the IADC guidelines is redressal of debris issues and invoking the modification serves to bring mega-constellations of satellites and every kind of satellite clearly within the ambit of the guidelines.

With regards to the definition of space debris in space law, the term has not been mentioned, defined or described in the space treaties. This could perhaps be attributed to the fact, that until the last space treaty (MA-79), space debris was never an issue. In fact, treaty negotiators were primarily concerned with which artificial objects should be considered as space objects and not with the effects of these objects after their active life.[50] In view of the fact that the term has not found its way into the space treaties for the past twenty-five years, it would be prudent to relegate the issue for long term resolution. For the short term, the IADC guidelines specifically define 'space debris' in Article 3.1 as:

> all man-made objects including fragments and elements thereof in Earth orbit or re-entering the atmosphere, that are non-functional.[51]

For the subject under focus, the definition is of space debris is presently suitable since it encompasses satellites of all kinds.

4.5.2 Declaring LEO as Limited Natural Resource

In specific terms, the IADC guidelines recognize the unique nature of LEO and GEO in Article 3.3.2 to ensure their future safe, sustainable use and further states that these regions should be protected regions with regards to generation of space debris. However, when it comes to the ITU Radio Regulations (RR), only the GEO finds mention. As a matter of fact, the term LEO finds no mention at all.

The ITU (RR) framework for space is based on the main principles of efficient use of and equitable access to the spectrum/orbit resources laid down in No. 196 of the ITU Constitution (Article 44), which stipulates that:

> In using frequency bands for radio services, Members shall bear in mind that radio frequencies and any associated orbits, including the geostationary-satellite orbit, are limited natural resources and that they must be used rationally, efficiently and economically, in conformity with the provisions of the Radio Regulations, so that countries or groups of countries may have equitable access to those orbits and frequencies, taking into account the special needs of the developing countries and the geographical situation of particular countries.[52]

---

[50] U.S. Congress, Office of Technology Assessment (1990).

[51] IADC Guidelines, *supra*, note 52.

[52] *Constitution and Convention of the International Telecommunication Union* 22 December 1992, Vol. 1825, 1–31251 (in force 1 July 1994) [ITU Constitution], Article 44.

Originally, the use of frequency bands for radio services in space was envisaged primarily for communication satellites in the GEO and hence the principle of efficient use and equitable access applied only to the GEO. The purpose of the proposed mega-constellations is also to use frequency bands and the orbit for the same general purpose of communications and hence there is but little reason why the frequency and orbital slot allocation in LEO should not be vested with the ITU. In keeping with the need of the present times as also the principle of rational, efficient and economic use the existing provision could be extended to include LEO thereby bringing the entire affair within the scope of Article 44.

The subject of frequency/orbital slot management of mega-constellations would ultimately rest with the ITU whose RR provisions are far more amenable to change than the space treaties. Towards this end, World radiocommunication conferences (WRC) are held every three to four years to review and revise the Radio Regulations. Revisions are made on the basis of an agenda determined by the ITU Council, which takes into account recommendations made by previous world radiocommunication conferences. The next WRC is scheduled for 2019 and while its agenda does contain Resolution 659 on "*Studies to accommodate requirements in the space operation service for non-geostationary satellites with short duration missions*",[53] a specific treatment of mega-constellations is conspicuous by its absence. The above study, if at all geared towards mega-constellations might not suffice because of the simple fact that by the time the study is completed and its recommendations implemented, entire constellations would be in place and post hoc solutions would be the only option.

The situation demands decisive action. As it is, frequency issues in case of the mega-constellations would fall within the ambit of ITU. Modifying the regulations to include LEO as a limited natural resource would enable far more rational, efficient and economic use. The WRC-19 could potentially make a huge headway in resolving the situation and enabling sustainable use of outer space for future generations.

4.5.3 *Non-functional Space Objects and Environmental Law*

As of August 2017, the number of satellites launched in Earth orbit are about 7500, of which only 1738 are functional and about 4635 are non-functional. These non-functional satellites are effectively space debris and add to the existing debris numbers.[54] Of these 1738 functional satellites, 1071 satellites are in LEO where the mega constellations are planned.[55] If all the constellations planned are launched,

[53] ITU, "WRC-19 Agenda and Relevant Resolutions", online: https://www.itu.int/oth/R1402000001/.

[54] As per ESA, as of January 2017, there are 29,000 objects bigger than 10 cm in size, 750,000 objects from 1 to 10 cm and 166 million objects from 1 mm to 1 cm. Ref ESA "Space Debris by Numbers", online: https://www.esa.int/Our_Activities/Operations/Space_Debris/Space_debris_by_the_numbers (accessed 11 February 2017).

[55] Union of Concerned Scientists, *UCS Satellite Database*, online: https://www.ucsusa.org/nuclear-weapons/space-weapons/satellite-database#.Wx6A4YpKjIV (accessed 29 July 2018).

this would straightaway result in a ten-fold increase in the population in LEO. An increased population would soon translate to thousands of non-functional objects in LEO rendering it unavailable unless the trend in abandoning non-functional objects in space is addressed.

The OST addresses the trend to a limited extent in that Article VI of the OST requires authorization and continued supervision of the space object by the appropriate state party. Also, Article VII makes it clear that the state which registers the space object shall retain the jurisdiction and control over such object while in space. From the foregoing, it is clear that rendering of a satellite as non-functional does not allow for abandonment. It does not absolve a state of its duties and responsibilities. This is further emphasized by Article IX[56] which states that the use of space shall be guided by the principle of co-operation and mutual assistance and all activities shall be conducted with due regard to corresponding interests of all other state parties. However, these provisions are general in nature and consequently, at the end of their useful life satellites continue to be abandoned in space. The situation with respect to mega-satellites is critical because of the large numbers involved and hence it would be essential to invoke principles of international law beyond the OST that are less prescriptive, more consent-based and effective.

The issue relates to the environment of space and the principles of international environmental law provide a helpful framework to address issues related to environmental degradation, particularly non-functional small satellites which are inherent in mega constellations. There exists general agreement on the hazards of the activity but no scientific certainty and the precautionary principle squarely addresses the question of how we ought to proceed in the face of unavoidable uncertainty and hence solutions are explored in this principle.

Principle 15 of the Rio declaration aims at avoiding the occurrence of serious or irreversible environmental damage, by demanding states to take precautionary measures to preserve the environment even when scientific evidence is inconclusive or where potential adverse effects of an activity are not fully understood. A similar manifestation of the spirit of this legal principle is manifest in Article IX of the OST in the 'harmful contamination' clause. It has been argued that the clause relates to inter-planetary exploration and back-contamination of earth, however, the general purpose of maintaining the sanctity of the space environment is amply manifest in this clause. The means are different but the ends desired are the same. The application of the principle has always been desired.

What is problematic in application of the above principles is the verification aspect; the aspect of ascertaining proof which is an issue in view of the limited capabilities for space object tracking. Here, it needs to be borne in mind that an emerging view of the precautionary principle is that it should be utilized to impose a reverse burden of proof in order to surpass the evidentiary difficulty inherent in proving the presence of environmentally harmful activities in space.[57] In this approach, a state interested in undertaking any space activity bears the onus of proving that such activities will

[56]OST, *supra* note 24, Article IX.

[57]Garnett and Parson (2016), Stacey (2016).

not produce adverse environmental consequences. The underlying rationale of the Precautionary Principle is that the producer of the alleged environmental damage is invariably placed in the best position to produce all the relevant information on its ongoing activities. With regards to constellation control, station keeping, maneuver and space traffic management, those operating or using the mega-constellations to provide services are best positioned to produce all the relevant information on its activities. Secondly, information on deorbit capability like operational parameters and reserve fuel would be available with the operator. Applying the Precautionary Principle in space ensures that the controlling state is responsible for providing proof that its activities are not degrading the environment of space. In the specific case of mega-constellations, the prime conceivable method of avoiding further environmental degradation would be of possessing enough deorbit capability once useful life of satellites nears an end. Invoking the obligation of 'avoiding harmful contamination' by deorbiting prior to completion of useful life of the satellite may go a long way to ensure the environment is free of non-functional satellites. The other methods of avoiding contamination by passivation, break-ups and other factors are already covered by the IADC guidelines and blame could be apportioned based on adherence/non-adherence to the guidelines.

4.5.4 Elevating UNCOPUOS Guidelines to Standards

The regulations in space draw on analogies in air law in terminology as also principles. The interlinking is only bound to increase as aeroplanes evolve to aerospace planes, air ports give way to aerospace ports and air traffic management expands to aerospace management. For quite some time, this has led to stringent calls for the equivalent of an ICAO in space to deal better with the multitude of issues there. The rationale, arguments put forth in advocating an ICAO for space are yet to be disputed in any manner and yet the possibility of such an international organisation continues to be remote. Regardless of the hurdles in transplanting aviation organisation onto space, the manner in which aviation regulations deal with common challenges, vulnerabilities faced by operators for a safe, efficient use of the environment hold many analogous prescriptions for space. Firstly, as in case of ICAO, the problem addressed is one faced primarily by civil rather than military agencies. The convergence is on civil users obtaining a common platform for prescriptions related to safe, secure, efficient and optimal utilization of space assets. The end result is commercial rather than military advantage. With regards to mega-constellation, the competition at worst would be limited to driving another player out of the market, buying out others rather than actually destroying the satellites. Such an environment would be fertile grounds for exploring implantation of ideas that work successfully in aviation. In light of the above, the possibility of implanting the concept of Standards and Recommended Practices (SARPs) of ICAO is explored.

The ICAO Council under the authority of 37 and 54 of the Chicago Convention adopts SARPs on issues of safety and efficiency of air navigation and designates them as Annexes to the Chicago Convention. In the SARPs, a "*Standard*" is a specification for physical characteristics, configuration, material, personnel or procedure, the uniform application of which is recognized as necessary for the safety or regularity of international air navigation and to which the Contracting States will confirm in accordance with the Convention.[58] On the other hand, a "Recommended Practice" is any specification for physical characteristics, …which is recognized as desirable in the interest of safety, regularity or efficiency of international air navigation and to which the Contracting States will endeavor to confirm in accordance with the Convention.[59]

Herein, a standard carries a recognition of necessity and hence becomes an obligation whereas a recommended practice is desirable and states only need to endeavor to confirm to it. Further, there are no strong penalties for not adhering to SARPs. States which cannot or do not wish to comply with SARPs are entitled to inform ICAO of non-adherence without risk of forfeiting any rights conferred by the Convention. Thus, in effect, the SARP is a mixed regulation with components of both hard and soft law which is followed successfully in the complex world of international aviation. The commercial interests in aviation are equally, if not more competitive and yet a convergence of interest on the common challenges of safety and optimal utilization of the environment enables adherence to regulations by free consent.

On similar lines are the IADC guidelines which have been developed by consent, are non-binding and were first published in 2002 and subsequently updated in 2007. These guidelines formed the basis for development of the UNCOPUOS Space Debris Mitigation guidelines which were endorsed by the UN General Assembly in its resolution 62/217 dated December 2007.[60] For mega-constellations, an IADC statement has been issued in 2017 which carries a range of recommendation akin to the SARPs in aviation. What is proposed is that the Legal Sub-Committee of the UNCOPUOS under the authority of Article III (for invocation of the environmental law principles) and Article IX of OST elevate these guidelines to the status of SARPs. The IADC guidelines which are being followed by many states since 2002 may be accorded the status of Standards based on treaty obligations of Articles II and IX as also custom evident since the past decade. The contentious issues may be treated as Recommended Practices which following modification could later gain the position of Standards. The above is not likely to resolve the issue in its entirety but may be considered as amongst a series of steps essential to handle emergent issues.

---

[58] *Consolidated Statement of Continuing ICAO Policies and Associated Practices Related Specifically to Air Navigation*, Assembly Res A36-13, Appendix A, ICAO Doc 9902 (2007).

[59] *Ibid.*

[60] UNCOPUOS "Debris Mitigation Guidelines", online: http://www.unoosa.org/pdf/publications/st_space_49E.pdf (accessed 29 July 2018).

4.6 Conclusion

The surge in small satellites presents a huge challenge in the short term and an equally huge opportunity for the future. Therefore, the future space utilization would depend on the dynamism of law to handle this surge which is unprecedented. Consequently, there is an emergent need to seek ingenious solutions which enable convergence of common interest, standardization of operations, conservation of the environment, dispute settlement and streamlining of methods of optimal utilization among other things. These needs can be met by following a practical, balanced approach that is neither too conservative nor drastic. Towards this end, a beginning can be made by agreeing on standard definitions for safety of operations, by solidifying practices of the IADC guidelines that have progressively gained universal acceptance into standards, by declaring the operating environment of LEO as a limited natural resource, by modifying ITU procedures and invoking precepts of environmental law which enable sustainable use and development of space. Significant gains for all of humanity for generations could be frittered away by leaving the domain unregulated and incorporating measures post hoc. As a matter of fact, in the eventuality of a debilitating collision in the proposed orbit, the entire belt may be rendered unusable and the possibility of post hoc action may be foregone entirely. The need of the hour is for decisive action which serves the purpose for generations to come and the recommendations prescribed are but the initial steps towards this end.

References

AMSAT-UK (2014) Successful launch of Kicksat carrying 104 sprite satellites. https://amsat-uk.org/2014/04/18/successful-launch-of-kicksat-carrying-104-sprite-satellites/. Accessed 2 May 2018 (18 April)

Brodkin J (2018) FCC approves space X plan to launch 4426 broadband satellites. arsTechnica. https://arstechnica.com/.../spacex-gets-fcc-approval-to-build-worldwide-satellite-broadband-network/. Accessed 4 Apr 2018 (30 March)

Clark S (2013) Silo-launched Dnepr rocket delivers 32 satellites to space. Space.Com. https://www.space.com/23738-dnepr-rocket-launches-32-satellites.html. Accessed 2 May 2018 (25 November)

Cummings WD (2014) Small satellites: advancing university scientific research and workforce development. Presentation to the scientific and technical subcommittee of the UN Committee on the Peaceful Uses of Outer Space, Vienna, 11 February 2014. http://www.unoosa.org/pdf/pres/stsc2014/tech-03E.pdf. Accessed 13 July 2018

David L (2017) Sweating the small stuff: cube sats swarm earth orbit. Scientific American. https://www.scientificamerican.com/article/sweating-the-small-stuff-cubesats-swarm-earth-orbit/. Accessed 16 Apr 2018 (12 July)

de Selding PB (2015) Signs of a satellite internet gold rush in burst of ITU filings. Space News. http://spacenews.com/signs-of-satellite-internet-gold-rush/. Accessed 4 Apr 2018 (23 January)

Freeland S (2015) A delicate balance: regulating microsatellite technology in a big satellite world. Univ West Syd Law Rev 18:1

Garnett K, Parson D (2016) Multi-case review of the application of EU law and case law. Risk Anal J (18 May)

IADC Steering Group (2017a) IADC statement on large constellations of satellites in low earth orbit, p 4. IADC-15-03. https://www.iadc-online.org/Documents/IADC%20Statement%20on%20Large%20Constellations%20rev%203.pdf. Accessed 27 July 2018 (September 2017)

IADC Steering Group (2017b) IADC statement on large constellation of satellites in LEO, IADC-15-03. https://www.iadc-online.org/Documents/IADC%20Statement%20on%20Large%20Constellations%20rev%203.pdf. Accessed 28 July 2018 (September 2017)

Grush L (2018a) Space X just launched two of its space internet satellites-the first of nearly 12,000. The Verge. https://www.theverge.com/2018/2/15/17016208/spacex-falcon-9-launch-starlink-microsat-2a-2b-paz-watch-live. Accessed 4 Apr 2018 (22 February)

Grush L (2018b) Company that launched satellite without permission gets new license to launch more probes. Verge. https://www.theverge.com/2018/10/4/17928452/swarm-technologies-spacebees-satellites-spacex-falcon-9-fcc-license. Accessed 7 Dec 2018 (4 October)

Jakhu RS (2016) Regulatory process for communication satellite frequency allocations. In: Pelton JN et al (eds) Handbook of satellite applications. Springer, New York

Jakhu RS, Pelton JN (2014) Small satellites and their regulation. Springer, New York, pp 53–54

Jakhu R, Sgobba T, Dempsey PS (eds) (2011) The need for an integrated regulatory regime for aviation and space. Springer, Vienna, pp 27–31

Jakhu R, Nyompong YOM, Sgobba T (2017) Regulatory framework and organisation for space debris removal and on orbit servicing of satellites. J Space Saf Eng 4(3–4):129

Larbi B et al (2017) Active debris removal for mega constellations: CubeSat possible? Paper presented at the 9th international workshop on satellite constellations and formation flying, Boulder, CO, USA, 19–21 June 2017

Lyall F, Larsen PB (2017) Space law: a treatise. Routledge, New York, p 249

Mabrouk E (2015) What are small sats and cube sats. https://www.nasa.gov/content/what-are-smallsats-and-cubesats. Accessed 10 Jan 2017 (26 February)

Masson-Zwaan T (2016) Registration of small satellites and the case of Netherlands. In: Marboe I (ed) Small satellites: regulatory challenges and changes. Brill, Leiden

Masson-Zwan T (2012) Operating under the radar: nano satellites, international obligations and national space laws. In: 55 Proceedings of IISL, p 566

NASA (2014) Orbital-2 mission to the international space station. NASA Media Press Kit. https://www.nasa.gov/sites/default/files/files/Orb2_PRESS_KIT.pdf. Accessed 2 May 2018 (July 2014)

Ohnishi M (2018) IADC: an overview of annual activities. Presentation at the 55th Session of STSC, UNCOPUOS. http://www.unoosa.org/documents/pdf/copuos/stsc/2018/tech-04E.pdf. Accessed 27 July 2018 (09 February)

Palkovitz N, Zwan TM (2015) Small but on the radar: the regulatory evolution of small satellites in the Netherlands. In: 58 Proceedings of IISL, p 601

Sheetz M (2018) Former Google engineers start up slammed by FCC for unauthorized sat launch. CNBC. https://www.cnbc.com/2018/03/09/swarm-technologies-slammed-by-fcc-for-unauthorized-satellite-launch.html. Accessed 16 July 2018 (09 March)

Stacey J (2016) Preventive justice, the precautionary principle and the rule of law. In: Tulich T et al (eds) Regulating preventive justice. Routledge, New York

Tallis J (2015) Remediating space debris: legal and technical barriers. Strat Stud Q 9(1): 86, 92–95

U.S. Congress, Office of Technology Assessment (1990) Orbiting debris: a space environmental problem. Background Paper, OTA-BP-ISC-72. U.S. Government Printing Office, Washington D.C., p 27. https://www.princeton.edu/~ota/disk2/1990/9033/903307.PDF. Accessed 30 July 2018

Venturini CC (2017) Improving mission success of cubesats. NASA Aerospace Report No. TOR-2017-01689. https://www.nasa.gov/…/improving_mission_success_of_cubesats_-_tor-2017-01689. Accessed 13 June 2018 (June 2017)

Walker R et al (2001) Analysis of effectiveness of space debris mitigation measures using the delta model. Adv Space Res 28(9):1437

Wall M (2013) Record setting rocket launch on Nov 19: the 29 satellites. Space.Com. https://www.space.com/23646-ors3-rocket-launch-satellites-description.html. Accessed 19 Apr 2018 (19 November)

Whitman R (2017) Smallest ever working satellites reach orbit. Extreme Tech. https://www.extremetech.com/extreme/253167-smallest-ever-working-satellite-size-postage-stamp. Accessed 2 May 2018 (27 July)

Chapter 5
Conclusion and Summary

Abstract The law relating to space has not changed drastically, what has changed is the context and this drives the need to adapt for contextual applicability. The broad principles of international space law were meant to guide future space exploration and utilization. Consequently, the law by design, is resilient to change and this resilience enables contextual adaptation. This thesis advocates viewing law in abstract, non-traditional terms for application to contemporary issues. Accordingly, a broad based, contemporary interpretation of the principle of peaceful use of space is advocated to envelope the wide variety of challenges to peace inherent in space today. This wide interpretation enables application of the precepts of law of other analogous legal disciplines that enable comprehensive regulation and foster greater international cooperation.

The foregoing thesis aims to be pragmatic and purposeful. It is hence rooted in the existing reality that the possibility of creating a new space regime to deal with the emergent issues in outer space is remote. No new space treaty has come about since the Moon Agreement of 1979 and hence a practical approach of seeking solutions within the existing scope of international law is undertaken. The international law relating to space has not changed drastically, what has changed is the context and this drives the need to perceive international law unconventionally.

Law is not meant to be static and this thesis on examining the evolutionary history and scholarship of the law related to space arrives at the conclusion that the existing space regime is inherently broad based and dynamic. The broad principles of international law were meant to guide the future space exploration and utilization. Consequently, the law by design, is resilient to change and this resilience enables contextual adaptation. This thesis advocates viewing law in abstract, non-traditional terms for application to contemporary issues.

Towards this end, the first chapter serves to introduce the scope of existing law to deal with the multitude of emergent issues and actors in outer space today. It draws attention to the changed context and the expanding scope of international law to address a variety of non-traditional issues ranging from rising cooperation in commercial space activities to a convergence of interests on common vulnerabilities like

K. K. Nair, *Small Satellites and Sustainable Development - Solutions in International Space Law*, SpringerBriefs in Law, https://doi.org/10.1007/978-3-030-18620-3_5

space debris. It summarizes that the changed context provides the right opportunity to shift focus from containing conflict to encouraging convergence in space.

The following chapter deals with the principle of the 'peaceful uses of outer space' which is the cornerstone of the space treaties. It examines the evolution and subsequent convoluted interpretation of the term by an examination of various factors ranging from the *travaux préparatoires* to state practice and arrives at the conclusion that the principle was never intended to be confined to a narrow military context or only to states. This chapter refutes and rejects traditional notions and points out that though the selective interpretation of peaceful use was to contain the military issues then prevailing, nothing stops an ordinary, expansive interpretation of the same for a contextual application. A broad-based interpretation of the term, as originally envisaged, enables wider application of international law to the contemporary issues of regulating commerce, unfair competition in space, regulating the surge in mega-constellations of small satellites, and enforcing space debris management measures for a sustainable, peaceful environment in space.

The third chapter examines the role of international law in dealing with the complexities of commercial space cooperation and in promoting harmonious advancement of converging interests. It demonstrates that the scope has expanded from containing conflict to promoting commercial cooperation today. Based on scholarly debates, case studies of the ISS, state models of cooperation and empirical data it arrives at the conclusion that the existing cooperative mechanisms are resilient, they work and enable a convergence of interests, but this convergence is a convergence of self-interest which is contrary to the common-interest principle in the space treaties. This demands tempering by suitable regulations. The common factor binding diverse interests in space is a need to conduct affairs in space peaceably and leveraging on this, an attempt is made to contextualize commercial cooperation under the broader rubric of peaceful use. By doing so, the general obligations of cooperation can be combined with specific obligations of peaceful use inherent in the space treaties. This, for the interim provides options for harmonious advancement and also ensures sustainable development of space. The ideal recourse would be to frame new international agreements, but that is a long-term solution.

In the fourth chapter, the emergent issues of mega-constellations of small satellites, consequent aspects of space debris and sustainable development of space are dealt with. These issues are common and all mankind is vulnerable to its impact. Therefore, a loose convergence on addressing these vulnerabilities already exists. An examination of the role of international law in progressing this convergence is undertaken. In doing so, firstly, recourse is taken to empirical data to establish the extent of the problem and assess the need for quick solutions. The conclusion arrived at is that the issue is both large in extent and critical. This is followed by a critical appraisal of the space treaties and ITU regulations which reveal many areas of inadequacy in dealing with the issue. A wide interpretation and application of the concept of peaceful use is helpful but does not resolve issues. Consequently, through Article III of OST-67, the precepts of international law are evoked. This opens the door to innovative solutions beyond the confines of the space treaties in a variety of areas including Regulations of the International Civil Aviation Organisation (ICAO) and precepts of international environmental law. The recommendations *de lega ferenda* include standardization of definitions to ensure a common platform, a declaration of the Low Earth Orbit as a limited natural resource, elevation of IADC guidelines to the level of 'Standards' as in ICAO regulations and application of the precepts of international environmental law, particularly the 'Precautionary Principle'. The discussion and analysis here lead to the conclusion that these areas of international law have been effective in dealing with common vulnerabilities in their respective domains and there exists little reason to not test their applicability in the domain of space.

Made in the USA
Monee, IL
16 September 2026

63227618R00057